The Cockpit Companion

Flight Assistance From the Right Seat

by Gordon Groene and Janet Groene

Jones Publishing, Inc.
N7450 Aanstad Road
P.O. Box 5000
Iola, WI 54945
Phone: 715-445-5000
Fax: 715-445-4053

Cockpit Companion
Flight Assistance From the Right Seat

by Gordon Groene and Janet Groene

Publisher/Acquisitions Editor:
Gregory Bayer

Editors:
Frank Hamilton, Julie King

Art Director/Cover Design:
Joni Clarke

Cover Photograph
Frank Hamilton

Production Team:
Jean Adams, Cindy Boutwell, Cindy McCarville

Published by: **Jones Publishing, Incorporated**
N7450 Aanstad Road
P.O. Box 5000
Iola, WI 54945
Phone: 715-445-5000
Fax: 715-445-4053

Copyright © 1995 by **Jones Publishing, Inc.** All rights reserved. No part of this book may be reproduced or transmitted in any form or by any means, electronic or mechanical, including photocopying, recording or by any information storage or retrieval systems without written permission from the publisher, except for the inclusion of brief quotations in a review.

This book was written to entertain aviation enthusiasts. The authors, publisher, editors and other staff members shall have neither liability or responsibility to any person or entity with respect to any loss or damage caused, or alleged to be caused, directly or indirectly, by the information contained in this book.

If you do not wish to be bound by the above, you may return this book to the publisher for a full refund.

This book is available at special quantity discounts for bulk purchases. For prices or other information, contact Greg Bayer at the above address.

10 9 8 7 6 5 4 3 2 1 Printed in the United States

ISBN 1-879825-22-8 $14.95 U.S. Funds
Library of Congress Catalog Card Number: **95-78205**

About the Authors

Janet and Gordon Groene
Photo by Gordon Groene

From the age of four, Gordon Groene knew that aviation would be his career. As a high school student in Berea, Ohio he was president of the model airplane club and he worked at area airports to pay for his first flight lessons. By the age of 16 he had a Private license. After serving two years in the Army in Korea, he earned Commercial, Instrument, Instructor, Multi-Engine and Airline Transport Pilot ratings.

His aviation career included flying for Youngstown Airways, National Distillers, and finally for a division of Gulf + Western.

At the age of 38, Groene "retired" from aviation to pursue a life at sea with Janet aboard their sailboat in the tropics. The couple earned their way by writing books and magazine articles about their adventures and, after ten happily homeless years on the go by sailboat and motor home, settled down in Florida where Gordon added a Cessna 310 to their stable. He is a 10,000-hour pilot.

Janet Groene began her writing career as a junior high student with her own column in the Berea News. By the time she was a college freshman, she had her own column in the

Cleveland Plain Dealer. Her books include *Cooking on the Go, How to Live Aboard a Boat,* and *Dressing Ship,* all published by Hearst Marine Books; *Living Aboard Your RV* and *Cooking Aboard Your RV,* published by Ragged Mountain/McGraw-Hill, and a number of travel guidebooks as well as countless magazine and newspaper features. The Groenes have won many awards for their writings, including the NMMA Directors' Award for boating journalism. Their jointly bylined books also include *Florida Under Sail, Natural Wonders of Ohio, Natural Wonders of Georgia, Country Roads of Ohio,* and *52 Florida Weekends,* all published by Country Roads Press, and *The ABCs of Boat Camping* (Sheridan House).

Dedication

To the memory of
Bonnie Gann and Archie Gann
1945-1993   1937-1993

Bonnie Gann began flying in 1973 and was the first woman to receive a private pilot license from the Key West Navy Flying Club. She held ASMEL, Commercial, Instrument, CFI, CFII and AGI ratings and was working on her ATP. She raced in more than 30 aviation events, always finishing in the top ten. She was active in the 99's, Florida Race Pilots Association, AOPA, Florida Grasshoppers, Baron Owners and many other groups.

While working on his instrument ground course, Archie Gann fell in love with Bonnie, his flight instructor. They married on February 14, 1985.

The couple flew away together January 16, 1993.

The Groenes will donate 5 percent of their royalties from this book to the Bonnie and Archie Gann Memorial Scholarship Fund, which annually awards scholarships to pilots to assist in obtaining their CFI Rating.

Table of Contents

Introduction 1

Part One: Getting It Up (The Joy of Flying Right Seat)
  1. Say Hello to Your Seat 3
  2. Alphabet Soup 7
  3. The Good News About Flying 19
  4. Let's Start With Weather 29
  5. What's in a Name .35
  6. Panel Discussion: Knobs, Gauges and Thingums 43
  7. What's Going On Here? 57

Part Two: Riding Shotgun (Without Shooting
          Yourself in the Foot)
  8. Health and Safety Aboard 89
  9. The Knacks of Snacks 103
  10. Starring Roles for the Right Seat 113
  11. Kids Aboard: Piloting for the Pint Sized 123

Part Three: Where To Get Off (Fly Me To the Moon, or
          Perhaps to Cleveland?)
  12. Great Places To Go In Your Airplane 141
      Amelia Island Plantation, Fernandina Beach, Florida .142
      The Bahamas 143
      Chalet Suzanne, Lake Wales, Florida 145
      Crested Butte, Colorado 146
      Downtown Chicago 148
      Downtown Cleveland 150
      Fort Scott, Kansas .151
      Jekyll Island, Georgia 152
      Monterey Peninsula, California 153
      Palm Springs, California 154
      Sugar Loaf Resort, Cedar, Michigan 156
      The Tides Inn, Irvington, Virginia 156

Index 159

Introduction

Dear Pilot,

As a Left Seater, you have at least a Private pilot's license and probably additional ratings too. We picture your airplane as a single or small twin, carrying from two to six people.

In the Right Seat, we see a non-pilot who is your spouse, best fishing buddy, parent, older child, or a business associate who travels with you often.

What can you, as a Left Seater, do to reassure the Right Seat and to enlist appropriate support while discouraging unwanted interference? What can you encourage the Right Seater to do, to make the flight more enjoyable for pilot and passengers?

Come join us for a Right Seat ride!

Dear Right Seater,

What you don't know can hurt you.

We once attended a seminar for non-pilots who were frequent Right Seaters in private planes. To our amazement, the instructors started out with how to check fuel, and were soon into how to make the airplane go up, down, faster and slower.

This was not, mind you, a Pinch Hitter course or ground school for student pilots. Its aim was to allay the fears of non-pilots by telling them how to fly a schoolroom chair. Many soon became goggle eyed with confusion. By the time the seminar ended, our guess is that most of them were thoroughly terrified at how much they didn't know, could not do, and did not remember from the fast-paced session.

We fled when the group broke up for a coffee break, so we'll never know just how far these well-meaning pilots went in encouraging Right Seaters to start grabbing controls and taking over the checklist. Their approach was in sharp contrast to what one corporate pilot tells the business travelers who sometimes ride Right Seat. "I'm only going to say this once. Don't touch anything." He says it without smiling, and he means it. We concur.

Our advice to any Right Seater who is expected (or even permitted) to drain the fuel sump, compute the gas load, or set the altimeter while the pilot does "more important" things, is to run screaming for the nearest exit. The Right Seat can play an important role in any flight, adding to its safety and comfort and fun, but pre-flight and piloting are specialized skills that should not be left to the unready.

The more you know about flying, the more active a role you'll want to take. For now, however, relax and let the pilot handle things just as he or she was trained to do alone. On the ground, you may take turns gassing up the car or checking the oil, but this is a different kettle of fish.

Vive la difference!

—Gordon Groene and Janet Groene

Chapter 1

Say Hello to Your Seat

All About Sex

Now that we have your attention, we'll admit that this is really a gender disclaimer. This book would have to be far longer if we inserted "he or she" with every reference to the pilot or passenger. Where gender-neutral terms won't work, we'll scatter the he's and she's randomly, and let you figure it out.

We know both males and females who are pilots, and both men and women who are dedicated Right Seaters and who have no intention of becoming pilots. Gender has nothing to do with who sits where.

Hello from the Left Seat

As a veteran of some 10,000 hours aloft, most of it in the Left Seat in single-pilot operations, I guess you'd say that I'm of the Buck Stops Here/Paddle Your Own Canoe/Crusty Curmudgeon school of thought when it comes to flying. While it's nice to have a Right Seater who loves flying as much as I do and who can take an active part in the trip, my first responsibility is to the flight. And that means a minimum of distractions, explanations, small talk, and unsolicited or inappropriate "help."

For some Right Seaters, small plane flying is only an occasional experience. As a practical matter, it wouldn't be worth the time and expense of becoming a licensed pilot or even a Pinch Hitter unless that person is motivated to take up flying as more than a sometime occurrence.

For other Right Seaters who ride in fast, multi-engine airplanes, even the merest Pinch Hitter taste of flying would require hours of instruction. Learning to fly would be pointless unless the person is serious about it and is in it for the long haul.

The point of this book is that the Right Seater can be a vital part of the flight — confidently, unafraid, and without learning to fly.

In this book, we hope to go into the important role of the non-flying Right Seater, with the accent on sense, safety, and the joy of flying.

—Gordon Groene

And from the Right Seat...

We Right Seaters divide into two camps. In the first are those who, like myself, aren't at all interested in the technical aspects of the flight. Under VFR conditions, I can enjoy the scenery. Views from airliners can't compare with those from low altitudes, and I've spent awed hours gazing out over mountains, shores, sagebrush and sere deserts, the azure brilliance of the Caribbean, and the lush green croplands of the American breadbasket.

In the clouds, however, I'm just as happy to read a good book or open my briefcase and get to work. Don't bother me with gauges and radios. Just get me there.

In the other camp are those future flyers, former flyers, and would-be pilots who don't want to miss a moment of the flying experience itself. Many of these will want to learn to chart a course, compute fuel use, put on a headset, and take the occasional turn at the stick.

They will want to go further than this book will take them. Those who want to be able to land the airplane if the pilot is disabled can take a Pinch Hitter course; others will go into flight training, and perhaps even into aviation careers for themselves.

For now, however, both of us have the same role in the flight: to support the Left Seat in every way possible without interfering with the flight, and to feast at the banquet of private flying.

—*Janet Groene*

Approach Control: How to Use This Book

Many times during the writing of this book, frustration turned to rage at our house, rage to giggles, giggles to tears, tears to frustration again. Only through some miracle were we able to keep the machetes sheathed as we wrangled over wording — Left Seat versus Right, pilot versus layman, tekkie versus English major.

As a pilot with full knowledge of all the subtleties, shadings, ramifications, limits and caveats of every facet of flying, Gordon wanted to be precise. And that sometimes meant droning on long after my eyes glazed over. Some compromises have to be made when we are trying to tell the Right Seat enough to be reassuring and informative, while being accurate about the basic facts, yet without blundering down byways and blind alleys of technical information that the Right Seat does not need to know nor care about.

Our theoretical Left Seater has a Private license and probably additional training and ratings. Our theoretical Right Seater is the pilot's spouse, parent, elder child, best fishing buddy, or business associate. While our theoretical aircraft is a four to six-place single engine airplane or light twin with a comprehensive equipment list, we've tried to keep in mind the full range of general aviation aircraft, from Piper Cubs to small jets.

The task seemed impossible as we rambled into one journalistic bramble patch after another. For example, most airplanes have starters, but for the reader who has an airplane that must be "propped," new rules apply. And if the propeller is constant speed or controllable pitch, things are different still. At some point, we had to draw the line at saying everything there is to say on a topic, in favor of keeping the Right Seater interested.

Whatever your aviation skill level, whatever your interest in flying, it's our hope that the Right Seat reader will barge through the first reading of this book without stopping at the boxed material. You may never want to know how ADF works, or how GPS manages its pseudo random code.

If looking at the instrument panel arouses your curiosity about these and other items, read further. If you want to know more than we have supplied, ask the pilot. And, if you want to know more than the pilot knows about how things work, write the education sources listed in Chapter 10 or ask a reference librarian to get you a book.

It is with this first, fast, read-through in mind that we placed a chapter on definitions in Chapter 6. Normally, a Glossary is tacked onto the back. However, aviation is a whole new language to the Right Seater, and one that is encountered even before arrival at the airport. It's part of the fun, the bond, and the brotherhood to understand at least part of the pilot's lingo.

We'd suggest that you speed through that chapter, "The ABCs of Flying," before getting into the rest of the book. You'll be pleased to find that you already know many of the abbreviations, acronyms and terms, which will be a confidence booster. Later, you can flip back to the ABCs as needed. In time, you'll feel right at home with the language, the poetry, and the nuts and bolts of flight.

Chapter 2

Alphabet Soup

Every industry has its jargon, so don't be daunted by a Left Seater who drops abbreviations by the dozen. A considerate pilot will, at least for the first reference, spell out the entire phrase, e.g. Air Traffic Control before throwing ATC at you, but some slips are bound to occur simply because some lingo is so common in aviation that the pilot thinks everybody knows what PTT or PIREPS mean.

Nobody can keep up with all the abbreviations used in all phases of aviation. Military pilots have their own patter. So do airline pilots, flying farmers, experimental aircraft builders, the space program, and all the aviation-related fields such as electronics, parachute jumping, and airport maintenance. Outsiders, even other pilots, are sometimes mystified by the resulting babble, so you're not alone.

Many airplanes are called by their abbreviated names (is MD-80 an airplane or a fuel additive or both?). And airports have a three-letter designation to eliminate any chance of confusion between, say DAY (Dayton) and DAB (Daytona Beach.) Such designators as LAX and JFK have come into common parlance but, unless you live there, you probably don't know that MCO is Orlando. In aviation shorthand, MCO also means maintenance carry-over item. It can get downright silly.

Alphabet Soup

Here are some common terms and abbreviations you'll hear as a Right Seater. For now, just scan them. Our guess is that you'll be pleased and surprised at how many you already know. Some of them may never be needed; others can be added to your vocabulary word by word, flight by flight, until you are a master at pilotspeak.

Abeam If you're a sailor, this one is easy. It means directly off to your port or starboard, at 90 degrees.

Abort To cancel a planned maneuver, usually a take-off.

AD An airworthiness directive, more commonly referred to as a bulletin. When a particular airplane or component shows a pattern of failure, the FAA alerts all owners of similar equipment. Usually, the AD specifies an inspection, repair, reinforcement or a replacement, and the time frame in which it must be done.

ADF Automatic direction finder.

ADIZ Air Defense Identification Zone, which cannot be penetrated without prior notice. It is clearly marked on charts.

Affirmative is pilotese for Yes; **Negative** is used instead of No. Clarity is paramount, so using these multisyllabic words leaves less room for error than Yup, Nah, and the like. Incidentally, contractions such as don't and can't are not used commonly in aviation, for the same reason. Do not, can not, etc. are usually spelled out.

Air pocket There's no such thing as a hole in the air, but you might experience CAT, or clear air turbulence. Usually, pilots refer to it simply as turbulence or, more succinctly, *&%#@.

Airway In CPR (which is always wise for a Right Seater to know) an airway is a breathing passage. In flying it's an invisible pathway defined by radio signals.

Alternate usually means alternate airport. Aviation's middle name is Redundancy, and its religion is Have An Ace in the Hole. In filing a flight plan, the pilot selects an alternate airport in case the airport of choice is closed (usually due to weather). The fuel load is calculated according to a worst case scenario, which would include en route delays, headwinds, detours around storms, and having to land at an alternate airport.

Altitude This one is tricky because it can refer to your altitude above ground level (AGL) or above mean sea level (MSL). See altimeters, page 44.

AOPA Aircraft Owners and Pilots Association, the voice of general aviation. Your Left Seater is probably a member.

APU Auxiliary power unit. Sometimes a portable unit is brought in to help start engines in cold weather.

ARTCC Air Route Traffic Control Center, usually referred to as The Center.

ATC Air traffic control.

ATIS Automatic Terminal Information Service. At airports where this recorded message is available, its frequency is noted on charts and the pilot is required to listen to it for basic local information before contacting a controller.

Box slang for transponder.

CAP Civil Air Patrol.

CAT Clear air turbulence, a bumpiness that sometimes catches you by surprise because there are no clouds in sight.

CAVU Ceiling and visibility unlimited.

CFR Crash fire rescue, a department at the airport.

CG Center of gravity, which the Left Seater will take into consideration when loading the airplane.

Clearance The meaning of the word is obvious, and you'll hear it often in many different references. Usually expressed as "You have clearance to..." or "We are cleared to..."

COMM Short for *communications*. If a runway or other service is in or out of *commission*, it's abbreviated DCMSND or COMSND on charts.

Crosswind Any wind that is not aligned with the runway can make landing tricky.

dB Noise, and there is plenty of it around airplanes, is measured in decibels. When you're shopping for headsets, their noise reduction ratings are expressed in dB.

DG Directional gyro, a compass.

DME Distance measuring equipment.

EAA Experimental Aircraft Association, a group that is responsible for the world-famous fly-ins at Oshkosh in the summer and Lakeland in the spring.

ELT Emergency Locator Transmitter, a radio transmitter attached to the aircraft structure which automatically transmits on 121.5 MHz and 243.0 MHz in the event of a crash. The ELT aids significantly in the search for missing aircraft.

ETA Estimated time of arrival. Used less often in conversation is ETD, or estimated time of departure.

FAA The Federal Aviation Administration.

FAR Federal air regulations.

FBO Unfortunately, there is no other term for the fixed base operator (the place at the airport where you buy fuel, use the john, check with FSS, and buy coffee). Sometimes it's called simply "The Operator".

FCC Radios must be licensed by the Federal Communications Commission and used within their guidelines, which is one of the best reasons for unlicensed Right Seaters to stay off the horn. Penalties for FCC violations start with a Warning Notice or, more severely, a Citation of Violation. A license may be suspended or revoked and in serious cases fines start at $100 and go as high as $10,000, a year in the slammer, or both.

Final approach Lined up with the runway and on the way in. Usually expressed as "on final."

FPM Feet per minute, used to express rate of climb or rate of descent.

Freq. Rhymes with peak. Frequency, as in "high freq" or "repeat the freq."

FSS Flight Service Station, the FAA facility that provides the pilot with weather information, NOTAMS, flight plans and other information.

GC Ground Control; GCA is a ground control approach, using radar.

General Aviation. That's us, as opposed to airline or military aviation.

GMT Aviation operates on Greenwich Mean Time, thus avoiding any confusion over Daylight Savings or Pacific Standard. Sometimes called **Zulu** time.

Go-Around To choose not to land as planned. Also called a **Missed Approach.** A common misconception for passengers or the Right Seat is that a go-around means that somebody goofed or that some emergency is in the making. Not to worry. A pilot might choose to go around simply because the previous airplane wasn't clearing the runway quickly enough.

GPS Global positioning system, a navigational system that is coming into widespread use.

Ground speed This is another of those tricky ones. You can be making 150 mph through the air and only 100 mph over the ground. This is caused by a meteorological phenomenon we call Groene's Law, i.e. no matter where you decide to fly, there is a headwind from that direction.

Holding pattern A specified pattern to be flown while the pilot is waiting for clearance to go elsewhere.

Homing beacon A low frequency, non-directional radio beacon the pilot reads with an ADF. Usually just called the Homer.

IFR Instrument (as opposed to visual) flight rules.

ILS Instrument landing system.

INOP Not working (inoperative).

International as in airport. Not necessarily a big jetport, it is any airport where Customs services are available. Flying to Canada, Mexico or the Bahamas requires stops at such an airport where you "clear," or do the paper work associated with leaving, or "enter," which is to complete customs and immigration formalities involved in entering or re-entering a country.

Knots Speed designation that refers to nautical miles per hour. Airspeed may be expressed as KIAS, or Knots Indicated Air Speed. Knots per hour is redundant; use it and you may be left DOA at the FBO.

Mayday Never used frivolously, this is a universal call for

The Cockpit Companion

help (M'aidez). It is used only when the threat is grave and immediate. In aviation, nobody jokes about bombs, hijacking, or Mayday.

Minimums Certain ceiling and visibility minimums apply to each airport and type of aircraft.

Missed approach See go-around.

Navaid Short for navigational aid, which could be anything from a homing beacon to VORTAC.

Niner A pilot's way of saying nine to avoid confusion with the word five.

NM Nautical Mile, or 6000 feet.

NOTAM Notice to airmen, for example a notice that a certain runway is closed or a radio facility is shut down.

Pattern Although the term can describe any route, such as a holding pattern, it is used most often to refer to landing sequence. Just as certain traffic practices govern right-of-way at unmarked intersections, aviation has its own way of sorting out who goes where, when. The busiest airport in the nation is not O'Hare with its fancy radar, radios, and armies of controllers. It's the relatively small field at Oshkosh during the EAA fly-in. Yet airplanes come and go in order.

PCA Positive control area. Aircraft can't fly into a PCA without permission from ATC.

PIREPS Pilot reports, in which one pilot shares information that will be helpful to other pilots headed for that same altitude, area, or destination.

Pitot tube A pipe that may be anything from a simple tube to a fancy device with built-in heaters, that sticks out into the air stream and measures ram air pressure. On the ground, it is usually kept covered to keep bugs and moisture out.

Port Sitting in the right seat, facing forward, port is on your left. **Starboard** is to your right.

Precip Precipitation, including rain, hail, snow, and sleet.

PTT Switch Push to talk, a microphone switch.

Ramp The area where you will fuel up, load the airplane or park, as opposed to a taxiway, where you will taxi, or a runway, where you will take off.

RON Remain Over Night, a notation used mostly by pilots who fly for hire.

Sectional Chart A chart in 1:500,000 scale (one inch equals 500,000 statute inches *or* 41,677 feet *or* approximately eight miles). If you want to go to a small airport in Florida, for example, you need to know whether it's found on the Miami, New Orleans, or Jacksonville Sectional.

SIGMET Significant meteorological information. These are major revelations that may mean a change of plans. Less severe reports are AIRMETS (Airmen's meteorological information).

Stall When speed isn't sufficient to supply enough lift to the wings, the airplane is said to "stall," even though the engine is making as much noise as it ever did. Yet, in an engine failure, the pilot can probably bring the plane in for a safe landing, in utter silence, without stalling. Don't try to figure it out.

SM Statute mile, or 5,280 feet.

Squawk The signal your airplane sends over a transponder that tells ATC who and where you are. ATC may, for example, tell the pilot to "Squawk 1330", a number that will be dialed into the transponder to identify you on the radar as a specific target.

The Cockpit Companion

The term **Squawk List** is also used in reference to a list that is made after each flight, naming any items that need to be checked or repaired. These squawks are also sometimes called **Writeups.**

TAC Terminal Area Chart, an enlargement of an area in which several airports are found, giving more detail than is found on Sectionals.

TCA Terminal Control Area, a positive control area requiring special equipment on the airplane, and certain pilot qualifications. It doesn't refer just to the area in the vicinity of the terminal on the ground, but to a widespread and carefully specified area of airspace.

Touch and go A quick way to practice landings and takeoffs without stopping in between (in other words to touch down and immediately take off).

Vector Not to be confused with Victor, this is a change of direction given to the pilot by ATC.

Victor Next to Roger he's one of the most popular chaps in aviation. It's short for VHF, and refers to airways based on Omni ranges.

VFR Visual flight rules; see and be seen. Usually it means being able to see the ground, but in clear air atop a bank of clouds, a pilot in the United States (but not always in other countries) can also fly legally VFR On Top by observing certain regulations. In short, not being able to see the ground doesn't necessarily mean that your non-instrument-rated Left Seater is in violation.

VORTAC An omni station co-located with a DME station plus a TACAN (military navigation) station.

Yoke The control column, which usually looks like a steering wheel, unless you have a **Stick**, which looks like a stick.

Alphabet Soup

The Phonetic Alphabet

Most of us can do a passable job of conveying a message phonetically, using designations off the top of the head. We might, for example, improvise by saying D as in Diaper. Big mistake. That's why an international phonetic alphabet has been devised to avoid any possibility of confusion. For instance, D as in Diaper could sound, over a static-filled transmission, like V as in Viper or P as in Piper.

In other countries where V is pronounced like B, or W like V, or V like F — well, you get the idea that the old system just had to go.

Here's the official alphabet used by aviators, sailors, and radio operators worldwide.

A Alfa	**J** Juliette	**S** Sierra
B Bravo	**K** Kilo	**T** Tango
C Charlie	**L** Lima	**U** Uniform
D Delta	**M** Mike	**V** Victor
E Echo	**N** November	**W** Whiskey
F Foxtrot	**O** Oscar	**X** X-Ray
G Golf	**P** Papa	**Y** Yankee
H Hotel	**Q** Quebec	**Z** Zulu
I India	**R** Romeo	

Silver Linings

Pilots use some terms more commonly than others, but it's likely you'll hear clouds referred to by their scientific names, such as nimbostratus or altocumulus, in addition to plainer

The Cockpit Companion

terms such as "fair weather cu" or "thunderstorms," or slang terms such as "mare's tails" or "mackerel sky."

It may be easier to sort out the terms if you remember the basic definitions for the five root words:

Alto, meaning high, refers to clouds at high altitude.

Cirrus, meaning curl, tuft, or hairlike fringe.

Cumulus, derived from the Latin word for heap or pile. Cumulus clouds are puffy.

Nimbus, based on the Latin word for cloud, in meteorology refers to a dark or threatening cloud.

Stratus, from the Latin meaning a strewing or covering, usually refers to horizontal or layered clouds.

From these basics, it is a tall step to learning what a cirrocumulus cloud looks like, or what a nimbostratus cloud means to the flight, or what altostratus clouds tell you about tomorrow's weather, but it's a start. To be honest, most pilots don't use these terms much in everyday flying; they're used more in the classroom than in the cockpit. We'll talk more about weather in chapter four.

Chapter 3

The Good News About Flying

Don't let anybody tell you it's just like driving a car. Flying is infinitely safer than highway travel. It's faster and more fun, the view is better, and the person in the Left Seat — even a low time private pilot — has had to prove his ability to fly and that he is physically fit to do so.

Imagine that! Your Left Seater has had more training than a taxi driver or even a school bus driver, and has had a recent medical check-up covering many things including hearing, eyesight, heart, and sugar level.

The airplane too has been under constant scrutiny from Day One, and has received maintenance even on things that ain't broke. Every turn of the screw has been signed off by a licensed mechanic. And, if that isn't enough, to warm the cockles of your heart, the airport — and all its runways and systems — has been inspected, maintained, and licensed to a fare-thee-well.

Before any airplane takes off, all three elements — pilot, aircraft and airport — must meet standards that are set by law.

One longtime pilot told us that his wife's greatest fear, when they began flying as a family, was that he'd be lost all the time because he has no sense of direction on the ground. If automobile drivers had all the resources available to sky drivers, they'd never be lost. Relax and enjoy the flight, confident that the pilot — no matter how inept a navigator on the road — has a whole armory of navigational ammunition including many things you won't see or hear, at least at first.

No other industry, hobby, or form of transportation is so finely regulated, examined, and inspected. Your pilot has had to jump through hoops to get that ticket, and it is one of her most treasured possessions, one of his proudest achievements. Aviation is an unforgiving club. Those who foul up are turned in not just by the feds, but by tower operators, ground personnel, and each other. The last person a pilot wants to meet up there is some blunderbuss who doesn't know his ship from shinola.

It never fails to amaze us that the same people who will hop into just any car anytime, with anybody, are scared skinny about riding in "those little" airplanes.

As Bruce Jenner, the Olympic gold medal winner and MU-2 pilot jokes, "It's not a small airplane. Just look at the payments."

Here, briefly summarized, is just part of what your pilot has had to do to get into that Left Seat:

All Pilots

All pilots must be at least 16 years old, in good health, and able to speak, read, and understand the English language. English is the international language of aviation; it's used in control towers worldwide. If a pilot cannot meet this requirement, a restricted certificate, which cannot be used in any flight requiring the use of English, is issued.

Incidentally, there is no upper age limit. As long as the pilot can pass the physical exam, the sky is the limit. In fact, older pilots — even older student pilots — have one of the best safety records in aviation. As the saying goes, there are old pilots and bold pilots, but no old, bold pilots.

Training is given in a variety of ways and you will hear pilots talking about these different methods. Most of the terms are self-explanatory. Ground School is classroom work. "Dual" (instruction) involves the pilot and a licensed instructor. The pilot's first solo (flight alone) is a happy milestone, usually involving a ceremony in which the student's shirttail is cut off.

After soloing, the pilot continues to log dual and solo hours,

Not all flight training is done in airplanes. Your Left Seater has probably had hours of experience at the controls, both real and simulated. (Photo courtesy of Embry-Riddle Aeronautical University)

including a solo cross-country (multi-leg course), hood time (the pilot's eyes are shrouded in such a way that the pilot must rely on instruments rather than visual clues from the ground or horizon), and perhaps some simulator time. Tests for each license include writtens, orals, and flight tests.

Pilots of all skill levels are also subjected to check rides (with a licensed instructor or examiner) and to a regular flight review at least every two years (the "bi-annual").

Student Pilot

A student pilot must be 16 and pass a Class III medical exam from a physician designated by the FAA (Federal Aviation Administration). Student pilots can not carry passengers, so you won't be flying Right Seat with a student unless you're the pilot in charge.

Recreational Pilot License

A sort of stepping-stone license was introduced a few years ago and it remains legal, if not common. Still, it's a fast track for the student pilot who wants to be able to make local flights with a passenger aboard, before getting a full-fledged private. The Recreational Pilot License requires a minimum of 15 hours dual and 15 hours solo, plus a Class III Medical. With such a license a pilot can take no more than one passenger, can fly no more than 50 nautical miles from the airport at which the pilot was checked out, and cannot use controlled airports.

Private Pilot

To have earned a Private Pilot ticket, your pilot must be at least 17 years old and have 35-40 hours of flight time (most people amass at least 55 hours before trying to pass the test) including some instrument and night flight training. The private pilot has passed a 60-question written exam, an oral exam and a test ride with an FAA examiner, and must pass a Third Class physical every other year.

Instrument Rating

If your Right Seater has an Instrument rating, it means at least 125 hours total time, 40 hours of instrument instruction, a written exam, and a check ride with an FAA examiner.

CFI (Certified Flight Instructor)

An instructor must be at least 18 years old, and have a Commercial or Airline Transport rating with an Instrument rating. This rating also requires a written exam and an FAA check ride.

Commercial Pilot

No one can fly for hire without earning a Commercial Pilot ticket. The applicant must be at least 18 years old, have a Class II medical (which must be renewed each year) and must have logged at least 250 hours. Again, there's a written test and an FAA check ride.

Commercial is, by the way, one of the most misused terms in aviation because so many people confuse "commercial pilot" with "airline pilot." All airline pilots are commercial pilots and much more, but not all commercial pilots fly for airlines.

Although the Left Seater who does not have a commercial license cannot charge you for the flight, it is permitted to let passengers pay for a pro rata share of the fuel and landing fees but not for other costs such as hangar rent and insurance. If there's any doubt, ask an attorney because bending the rules could land the pilot in hot water with the FAA. Worse still, his insurance carrier could refuse to pay any injury claims made by a Right Seater.

Multi-Engine Rating

Any Private or Commercial pilot can take additional dual for as long as is needed to pass an FAA check ride in a multi-engine airplane. You may see the abbreviation SMEL, which stands for Single-Multi-Engine-Land.

Airline Transport Pilot

The ATP (Airline Transport Pilot) has at least 1500 hours (1200 hours for rotorcraft), a Class I physical every six months, a lengthy written test, and a rigorous FAA check ride.

Other Ratings

Pilots must have special ratings to fly helicopters, seaplanes, gliders, hot air balloons, and so on.

Medical Requirements

Pilots usually refer to the examination as the FAA physical and the resulting certificate as the medical. The complete regimen as spelled out in the FAR (Federal Air Regulations) covers eight pages of fine print, and are available from the FAA Aeromedical Certification Division, AAM-300, Box 26080, Oklahoma City, OK 73126-5083.

We won't list all the requirements here, but basically the pilot has been checked, not just by any physician but an FAA AME (Authorized Medical Examiner) who is designated by, and responsible to, the FAA, for:

Third Class Medical Certificate (Class III)

Vision and eye health including the ability to distinguish aviation signal red, signal green and white, hearing and ear health, equilibrium, blood pressure and cardiovascular health, neurological health, and general health, with the expectation that the pilot can be reasonably expected, based on present information and health history, to maintain this state for the next two years.

Extensive forms must be filled out in which the pilot must report all medications being taken, any visits made to any health professionals (including mental health or substance abuse treatment) during the past three years or since his last FAA physical, and any substance dependence. The FAA also reserves the right to request information from the National Driver Register about the pilot's traffic violations involving drugs or alcohol, health history, and physical description.

To fail to report that one is undergoing psychiatric care or is taking, say, blood pressure medication, could invalidate the pilot's license, insurance, or both. To make a fraudulent statement on the medical application form is to risk a fine of $250,000 plus 5 years in the slammer. It is also a federal rap to falsify one's logbook or a medical certificate.

Second Class Medical Certificate (Class II)

All the above, plus tighter standards, such as somewhat better vision and hearing.

First Class Medical Certificate (Class I)

All the above plus normal color vision. Between the ages of 35 and 40 the applicant must have an electrocardiographic examination and after age 40 must show a degree of circulatory efficiency that is compatible with high altitude flight. Prescribed blood pressure limits must also be met and, in general, a closer look is taken at everything covered in Third and Second Class exams.

Waivers

Contrary to what many laymen think, a pilot's license can still be issued to people who wear glasses or contact lenses, or who have certain limitations such as a missing limb. Usually, waivers are considered on a case-by-case basis depending on the problem and the aircraft or its adaptive equipment. FAA doctors also have the authority to issue a certificate for a more limited time period than the customary two years.

Additional tests

Many pilots must meet far more than FAA-mandated physicals, or meet them more frequently than the FAA prescribes. Gordon, for example, was subjected by one employer to a battery of psychological tests, as well as the usual Class I physical.

The Cockpit Companion

Sometimes such additional tests are required by the employer, sometimes by the insurance company.

And There's More...

The foregoing are just a brief summary of the requirements met by the person in the Left Seat of an airplane, but it's only part of the story. The person in your Left Seat probably has far more time and training than these minimums. Additionally, pilots earn type ratings that are required to fly different kinds of aircraft, and most pilots also take periodic flight reviews.

Most pilots constantly upgrade and update their skills voluntarily via free programs, lectures, and seminars sponsored by the FAA, AOPA, and other groups. Too, additional training or higher minimum hours may be required by insurance companies, employers of commercial pilots, flying clubs, or airplane rental firms.

Take another look at that person in the Left Seat. You're looking at somebody special.

In no other industry are equipment, operators, and support systems so fully scrutinized and regulated.
(Photo courtesy of Beech Aircraft Corporation)

About Charter Flying

Let's talk about those times when you may fly in the Right Seat as a paying passenger with a pilot who is a stranger to you, and not a family member or friend. Pilots are people, and that means they have many different personalities and philosophies. Because charter flying is, in many cases, an entry level position, your pilot will have all the licenses and qualifications but perhaps not the self-confidence of a seasoned skipper.

So, play it by ear. You'll find some pilots talkative, others very reserved, still others relaxed and open to your communication. And some are impossible to read. If you do not have a headset and don't know what's happening over the radio, you may think you're being ignored or may otherwise get the wrong impression simply because the pilot is listening to a very involved set of instructions from controllers.

All of the above applies doubly to any helicopter in which you're riding shotgun (usually in the left seat) because the pilot is busier with a less stable, more complex machine.

Keep in mind too that there is often more than one right way to fly, and techniques vary according to the kind of training a pilot has had. For example, a long time Right Seater once pointed out to Gordon that he'd "forgotten" to turn on the boost pumps (fuel pumps that back up the engine-driven pumps) during the runup. However, it's Gordon's practice to turn off the boost pumps as soon as the engines are running smoothly, to verify that the engine pumps are operating. They aren't turned on again until he's in position for takeoff. Because he did something different from what that Right Seater was accustomed to seeing, the passenger thought he had goofed.

Aircraft Inspections

It is not just the pilot who is constantly under the microscope. Aircraft have to continue to meet rigorous standards too. It begins with the certification of the design — a step that can take years and cost a fortune. Manufacturers spend millions of dollars testing components, and even entire engines and aircraft,

The Cockpit Companion

to the point where they self-destruct.

Engineers then determine a generous margin of safety, and specify at what point in its life cycle each item must be inspected or replaced. A service history is maintained on every aircraft, and if a pattern of failure develops, all aircraft (or engines or components) of that type are assumed to be prone to the same failure under the same circumstances and all the other owners are notified.

Such directives, called AD's (Airworthiness Directives), or Bulletins, are fairly common but most of them are simple to correct, or will not be needed until some date far in the future, or both. In addition to this monitoring of the fleet as a whole, every airplane is required to have an annual inspection, signed for by an authorized inspector (AI). The AI is a licensed A&P (Aircraft and Powerplant) mechanic who has an additional FAA authorization to sign for this inspection. By signing, the AI puts a lifetime of experience, as well as a future career, on the line, so the inspection is meaningful indeed.

In addition, routine engine maintenance is specified by the engine manufacturer. Usually maintenance steps are taken every 25 or 50 hours, depending on how often the oil is to be changed. People who work on airplanes are curious, and are first rate detectives. While the oil is draining, it's almost a certainty that the mechanic is looking around for any other problems or irregularities. It's more than a matter of looking for things to be fixed, to boost his income. It's part of the pride package, the privilege, and the burden that goes with being a licensed aircraft mechanic.

Any airplane that is used for hire (and that includes any rented airplane you're flying in) has an additional inspection every 100 hours. It's the same as the annual, except that it can be signed off by any A&P, not necessarily an AI. Add to this the fact that the pilot knows it is his or her responsibility to report and repair any problem that would affect a flight, and you can see that there is more to flying than just jumping into Left Seat and letting rip.

The Pinch Hitter Course

Although Pinch Hitter training is best known as an AOPA course aimed at teaching the non-pilot enough to land the airplane in an emergency, anyone who wants to learn these skills can sign up for flying lessons at any flight school.

In an organized course, you may be given some sort of certificate; in an individual course you'll simply be a student pilot who may choose to quit as soon as you feel you could, under ideal conditions, bring the ideal airplane to a survivable landing if the pilot becomes incapacitated.

If the instructor knows this is your goal, emphasis will be placed on these basics as they apply to the kind of airplane in which you usually ride Right Seat. Without a ticket, you'll have no rights, but you will have skills that may provide peace of mind.

The advantage to one-on-one training is that you can take it when and where you choose, at your own pace. Group courses, such as those offered by AOPA, also have several advantages. For one, they may be taught in the Right Seat where you'll actually be sitting if you have to take over the controls, unlike actual flight training, in which you're put into the Left Seat. A group course is probably less expensive than a private, piecemeal course. And, it's usually given in connection with some other event, which makes it sociable and much more fun.

AOPA Pinch Hitter courses cost $250-$350, plus aircraft rental, for full instruction. A ground school only course will cost approximately $75. The AOPA Air Safety Foundation also offers booklets for those who want to read more about how to pinch hit. For information about courses or the booklets, call AOPA at (800) 638-3101.

In praise of those who take a Pinch Hitter course, we say, "Good for you!" And in defense of those who choose not to take up flight training, we point out that the chances of your being stranded Up There are miniscule. In both cases, we remind both Left Seat and Right that a little knowledge can be a dangerous thing.

Chapter 4

Let's Start with Weather

When it comes to weather, everybody is an expert. Aunt Minnie feels it in her bones. Uncle Buzz fishes by the tides and barometric pressures, so he watches the Weather Channel for hours on end. Maizie and Chet do a lot of airline travel on business, so they know about cold fronts and downbursts.

Unfortunately perhaps, weather is the aspect of flying that most Right Seaters feel most qualified to comment on, so it's also the aspect most likely to cause friction between pilot and passenger. Here are just two examples.

We know a woman who dislikes flying in bumpy weather. Her pilot-husband knows his abilities, his instruments, and his airplane, and he does not exceed safe limits. However in flying under safe, but bumpy conditions, he neglects one very important factor: passenger comfort.

In this family, flying turns into a power struggle. He's right in saying it is safe and prudent. She is right in thinking that a considerate pilot would fly earlier, later, higher, or whatever, to make her more comfortable.

In another case, that of a very powerful politician on a tight campaign schedule, weather didn't matter at all. The politician was willing to weather any discomfort as long as he was on time for his next speech. When his pilot refused to take off in dirty weather, the politician sacked him.

No matter what you think you know about weather, it's important that the pilot's decision be law. The pilot has access to better, fresher, and more complete meteorological information than the average person, and is using this information to make a great many decisions.

In short, the Right Seater can get an excellent weather picture from television to help in deciding whether to pack the galoshes, but the Left Seater has access to a much wider range of information that will bear on whether you go, what route will be followed and at what altitude, what airports will be used, how much fuel will be needed, and so on.

When you get a weather sequence from any Flight Service Station it's always given in a standard format, so the Right Seater can quickly learn to take it down in shorthand. It will tell you:

1. Name of reporting station
2. Ceiling
3. Visibility
4. Temperature
5. Dewpoint
6. Wind direction and speed
7. Altimeter setting (barometric pressure)

It will be lightning fast and will sound like this, "Daytona Beach, three thousand broken, ten miles and haze, temperature eighty-five, dewpoint eighty, wind one twenty at twelve, altimeter three zero zero five."

On the teletype, this same report would read, "DAB 30 BKN 10 H 85/80 120/12 30.05." So, if you're copying a weather sequence heard over the radio, you can help the pilot by learning to take it down this way.

At some point during the weather briefing, the pilot may also ask the exact time of official sunset or sunrise. They are different each day in each destination, and navigation lights must be used accordingly.

Here are some of the things that the pilot will be looking for during weather briefings:

*Clouds tell an eloquent story about
the present and future weather.*
(Photo courtesy of Embry-Riddle Aeronautical University)

Ceiling and Visibility

One set of regulations applies to Instrument Flight Rules (IFR) and another to Visual Flight Rules (VFR). Unless the ceiling (bottoms of the clouds) and visibility (how far you can see) meets or exceeds minimums, it's no go. Period.

In any weather briefing, you'll hear three terms used to describe clouds: scattered (a few clouds), broken (a lot of clouds) and overcast (solid cloud).

Let's say you're flying VFR over the entire route. The pilot will have to make sure that the weather is VFR at the airport where you'll take off, at the airport where you'll land, and any points between. That's just the beginning of the things to be looked at however, because you may be dealing with two different weather systems. Read on.

Winds

Because weather reporting stations are few and far between, the pilot has to be able to interpret reports received from stations that may be 50 miles or more apart. If both reporting stations are in the same air mass, generally the weather will be the same between the two.

However, if the winds at the destination are, say, northerly and are southeasterly at the departure point, the pilot can

If you're interested, the pilot will give you a complete picture of the weather the flight will encounter.
(Photo courtesy of Embry-Riddle University)

deduce that there's probably a front (a border line between two air masses) between the two reporting stations. So the Left Seater will want to probe deeper by studying previous reports and talking to the meteorologist to know exactly what will be encountered in passing from one system to the next.

Winds will also be studied for their speed and direction at various altitudes, so the pilot can compute fuel use and choose the fastest, most comfortable en route altitude.

Temperature and Dewpoint

During the weather briefing, the pilot will be told the temperature and the dewpoint (the temperature at which moisture condenses) at various reporting stations. Because this condensation is cloud or fog, it's an important indicator to the pilot. By knowing both temperature and dewpoint, she'll know where to expect the bases of the clouds. And, by noting temperature and dewpoint trends, he'll know whether the clouds are getting higher or lower. If the two numbers approach each other, fog can be expected on the surface.

Barometric Pressure

In addition to telling the pilot how to set the altimeter (more about that later), barometric pressures are indicators of weather trends. A rising barometer generally means improving conditions; falling pressure means deteriorating weather.

And, in IFR Flight

In addition to the input listed above, the IFR pilot will want preflight weather information about such things as:

Tops

By learning where the tops of the clouds are, the pilot can plan to fly in clear air above them.

Imbedded Thunderstorms

When flying in the clouds, especially without radar, the pilot needs to know which ones could contain thunderstorms, which must be avoided.

Alternate Airports

In each instrument flight, the pilot will name a destination airport and an alternate airport to be used if weather at the first airport is below minimum operating conditions. Weather at both airports will be checked in advance, and kept track of en route.

Freezing Level

In weather briefings the pilot will be told at what altitude the airplane will enter 32-degree air. From this, plus dewpoint information, it can be determined whether ice accumulation will be a problem.

Chapter 5

What's in a Name? Empennage, and Other Incomprehensible Terms

Fortunately, ***empennage*** (meaning the tail assembly) is not a word anybody uses much. It's in the textbooks, but that's about as far as it gets. It is useful, though, to know the names for other major parts of the aircraft, especially since some of them can be confusing.

When you say, for instance, "By golly, Ed, look at them gravel nicks on the spinner," it sounds much more laid back than saying, "By golly, Ed, look at them gravel nicks on the pointy thing on the nose."

The Wings

Although some airplanes have ***slats***, which are fixtures on the front ***(leading edge)*** of the wings, it's more likely you'll be looking at only two or three moveable surfaces on the ***trailing edge***, or rear, of the wings. Don't confuse the ***ailerons***, which are usually outboard, with the ***flaps***, which are usually closer to the fuselage.

Ailerons, you'll soon observe, work in opposition to each

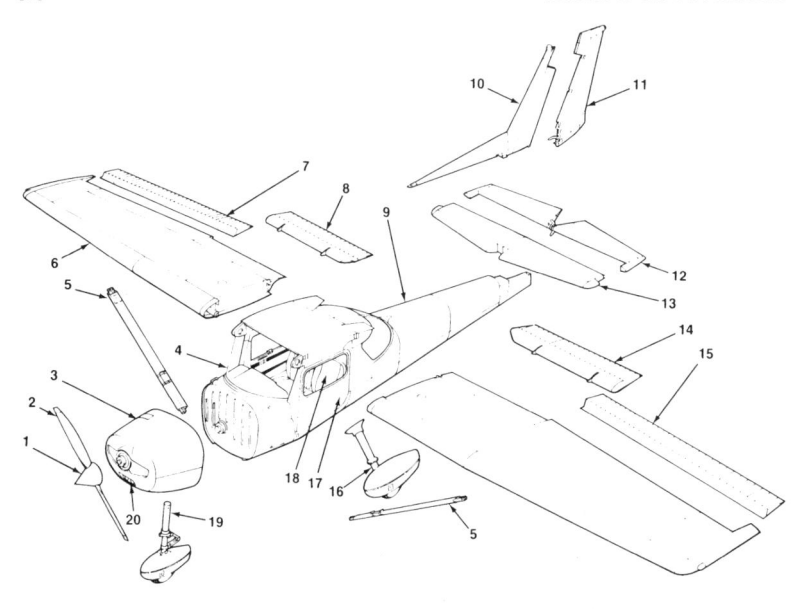

The Main Parts of an Airplane

1. Spinner
2. Propeller
3. Engine Cowl
4. Windshield
5. Wing Strut
6. Wing
7. Right Aileron
8. Right Flap
9. Fuselage
10. Vertical Stabilizer
11. Rudder
12. Elevator
13. Horizontal Stabilizer
14. Left Flap
15. Left Aileron
16. Main Landing Gear
17. Door
18. Seat
19. Nose Gear
20. Landing Lights

other. One goes up, the other goes down, creating an uneven air flow that makes one wing low, one high, and allowing the airplane to bank and turn.

The flaps are sometimes partially extended on takeoff to increase the curve of the wing — thus increasing lift. On landing they may be partially extended to increase lift during the low speed maneuvering required. On the final approach, they're put to their full "down" position to add drag to the lift and slow the airplane as it nears the runway.

Depending on the airplane, they may be left down now to aid deceleration on the runway, or retracted after landing to increase the effectiveness of the wheel brakes. If you're an avid Right Seater who wants to get involved, operating the flaps is basically a copilot function and may be one of the things the pilot asks you to handle.

On simpler airplanes, flaps are activated by a lever between the two front seats. It looks somewhat like an emergency brake handle in a sports car, and is equally handy to both seats.

On other airplanes the flaps, which may be electrically or hydraulically activated, are operated by a small handle on the instrument panel, handy to both Left and Right Seats. The handle is usually next to an indicator that shows what percentage of flap is extended.

Even if you've been Chief Flap Flipper for years, don't be insulted if the pilot snaps the job away from you in a hurry during a dicey go-around. At such times, the pilot will want to maintain complete "feel" for the lift.

Although you've seen **speed brakes**, or **spoilers**, on airliners, they're not common in small airplanes except on gliders.

It's probable that some or all of the fuel tankage is inside the wings, so you'll observe *fuel caps*, various **vents**, and **drains** on the bottom of the wings. If you see any fuel draining from any of them during flight, alert the pilot.

Little rope-like gizmos on the trailing edge are conductive wicks that bleed off any static charge that may build up on the airplane. They're called **static discharge wicks** or **static eliminators**.

Some wings have supports called **struts**. Others are supported entirely by unseen internal **spars**. In any case, don't be alarmed if the wings bend and flop a little. The flexibility is designed in, and is part of the strength of the wing rather than a weakness.

The Tail

The tail in most designs consists of **horizontal and vertical** (also called the *fin*) **stabilizers**, which do not move, and the **rudder** and **elevators**, which do. The rudder is mounted vertically and moves side to side. The horizontal parts that move are elevators, which move up and down.

Variations on this theme include the V-tail Bonanza, which combines the functions of elevator and rudder in only two sur-

The static eliminator wicks, clearly seen here trailing from the wing tips, bleed off static electricity that can build up.
(Photo courtesy of Beech Aircraft Corporation)

faces, and the stabilator, used on some Cessnas and Pipers, which is one large, moveable surface that acts as both stabilizer and elevator.

On the trailing edge of the elevator, you may see another small, hinged, moveable surface called a ***trim tab.*** It's operated from the cockpit by a trim wheel or lever and its purpose is to remove from the control wheel any load created by changes in CG (e.g. changing fuel load). It's very small but also very powerful, so like everything else, it gets a good once-over during the walk-around. Trim tabs might also be seen on the rudder or ailerons. If there is no trim tab on the elevator, the trim function is probably handled by a moveable stabilizer — a design commonly seen on jets.

Changing direction in an airplane, as you've already noticed, is not like hanging a 4-wheel drift in a car and pasting the passenger hard against the far door. When a turn is done correctly, passengers will not even know it if their eyes are closed.

To keep passengers solidly in their seats, the pilot uses the ailerons, rudder and elevators in perfect concert to keep the forces of gravity working at an angle, much as you do when you're whirling a bucket of water in a circle and none sloshes out.

If you haven't yet begun to ponder the kind of coordination needed to do all this while keeping the throttle(s) at an appropriate level, think about it now. As we've said before, that person in the Left Seat is somebody special.

The Nose

Up at the front end you'll find the *propeller*, its *hub* and the *spinner* that covers it, and the engine which is usually inside a *cowling*. In a *tricycle gear* airplane, the *nose gear* is here too. In front of the tire(s) you will see wedges, or *chocks*, which the pilot might ask you to remove. Usually they're just kicked out of the way and left for the next airplane that parks in this spot.

The important thing to remember about this end of the airplane is that the propeller can be deadly. Worse still, it's invisible when it's turning. Lesson One in hanging out at airports is to stay out of a propeller's plane of rotation. Always treat the propeller as though it is about to leap to life. Move in such a way that, if you lose your balance, you'll fall away from the prop.

A note about protocol. You don't yell Fire in a crowded theater without good reason. You don't yell Shark at the beach unless Jaws is actually there. And you don't joke at airports about Clear the Propeller. It's a heads-up phrase, not to be taken lightly, meaning that an engine is about to be started. If you're not clear, sound off and get clear pronto.

It's not likely that an engine will be started without these words, but there is always the possibility they won't be used, or that you won't hear them. So, as we said, treat the prop as though it's about to start, and don't give it a chance to bisect you.

We have more to say about the propeller in Chapter 7.

The Lights

You'll note a variety of lights on the wings, belly and tail of the plane. Because they sometimes reflect weirdly when you're in the clouds, fog, or any precip, it's a good idea to keep them in mind so you can reassure nervous passengers. A ***strobe*** (flashing) light, for instance, might bounce off a cloud and make you think at first that that lightning is striking all around, or a red light could be mistaken, at a quick glance, for a fire.

All aircraft that fly at night are required to have ***navigation lights***, a green light on the right wing, red light on the left wing, and a white light on the tail. In addition, the airplane will probably have a ***rotating beacon*** in red or white, mounted anywhere but often atop the tail, and/or clear white ***strobe lights***. Usually strobes are installed in sets of three, and are placed on the wing tips and tail or belly. Brilliant and flashing, they draw attention to the airplane, after which other pilots take a closer look at the red, white and green nav lights to get a clearer picture of the size and direction of the airplane.

If you make the walk-around check with the pilot, the two of you will take a close look at each of the nav lights and, if the flight will proceed into darkness, the lights will be checked under power to make sure they light. Here's where a second person can be helpful. You can observe the light while the pilot flips switches.

Under way, you can see the plastic ***telltales*** that reflect from the lights so you know the wingtip lights are on. Usually, the tail light can't be seen in flight.

Stuff Underneath and on Top

In addition to the ***main landing gear***, the airplane that does not have a nose gear will will have a little ***tail wheel*** or, in some antiques a ***tail skid***. A fixed gear stays where it is; if it disappears it's a ***retractable gear***.

Located around the aircraft, usually on the fuselage or the tail, are various ***antennas***. The communications antennas are usually straight whips about 30 inches long, and are generally mounted atop the fuselage.

The **VOR** (Very high frequency Omni Range, sometimes called Omni) **antenna** is two short whips mounted on a hori-

The Cockpit Companion

zontal plane high on the rudder. Another type of VOR antenna is two blades about 12 X 6 inches, mounted on either side of the vertical stabilizer. Another style is a pair of metal tubes, bent to about the same dimensions and resembling towel racks.

The **transponder antenna** is a small, tapered blade about 2-1/2 inches long usually installed on the belly. If the airplane has a DME, a second similar blade will be mounted nearby. If you see a teardrop shaped housing on the belly, it's probably the direction antenna for the ADF (Automatic Direction Finder). Some newer ones are small plastic blocks, sometimes mounted flush with the skin. Older ADF's also require a **sense antenna**, a long wire that runs from the top of the cabin to the tip of the vertical stabilizer.

A **Loran antenna** usually looks like a communications antenna. In a small airplane, a little whisker at the top of the windshield may be used as a **glide path antenna**. A dinky length of wire mounted somewhere towards the rear of the airplane is the **Emergency Locator Beacon antenna**.

On small airplanes, the **pitot** (rhymes with speedo) **tube** usually sticks out of one wing. A basic pitot tube is just a piece of pipe that is long enough to measure the pressure of the air before it is disturbed by the wing. The pressure is transferred to the airspeed indicator, where it works on a bellows that moves the hands on the dial and gives a direct readout in miles-per-hour or knots.

The pitot tube is usually kept covered when the airplane is on the ground because mud daubers love to nest inside. Don't ever blow into a pitot tube in an attempt to clear it. You'll ruin the airspeed indicator, and you may get a mouth full of wasps.

The cover for the pitot tube (usually it's a simple cloth envelope) is just one of the things that is removed before flight. Some items, such as the sun covers are hard to miss. Others, such as the **control locks**, are made hard to miss by the addition of flags or pennants. These locks are used to keep control surfaces from banging around in the breeze when the airplane is at rest. Obviously, the consequences of trying to take off with the controls locked are not pretty to contemplate. Thus the fancy reminders.

Chapter 6

Panel Discussion: Knobs, Gauges and Thingums

You'll spend endless hours staring at an expanse of gauges and dials, especially under instrument conditions when there's nothing else to look at. What at first seems like an overwhelm-

Beechcraft Bonanza A36 instrument panel
(Photo courtesy of the Beech Aircraft Corporation).

ing mess of monitors, all screaming for your attention, soon reveals itself to be somewhat simpler.

For one thing, some of the instruments may be duplicated, one set for each seat, so there aren't as many as you thought. For another, most of them are plain English readable, although you'll soon learn that some of them need some translation before they give you the whole story.

They can scare you shiftless if you jump to the wrong conclusions. Worse still, back seaters will be chewing your ear off wanting to know little things like how high are we and are we there yet, not to mention big things like "That fuel gauge reads Empty! Don't just sit there, send a Mayday!" So, the more you know about the eloquent story that the panel is telling to the Left Seat, the more confident you'll be and the more reassuring you can be to nervous passengers.

Flight Instruments

Altimeter

It's simply a very sensitive, expanded scale barometer that reads your altitude in feet above sea level if the pilot has first dialed in the local barometric pressure. It can tell you how high you are above the ground, but only if you know how high the ground is. So, If you started from, say, Miami at sea level, and flew to Denver at over 5000 feet without resetting the gauge, the altimeter would still read above 5000 feet instead of zero when you're sitting on the runway at Stapleton.

You'll see the pilot set the altimeter at the start of the flight, at each waypoint, and before landing. Altimeter setting/barometric pressure is one of the most important pieces of information that is fed constantly into the flight equation, and you'll hear it repeatedly.

Airspeed Indicator

Although the name would seem to indicate that this gauge tells how fast you're going through the air, it's another of those gauges that may not tell you exactly what you think it does.

The airspeed indicator measures the pressure of ram air in a

The Cockpit Companion

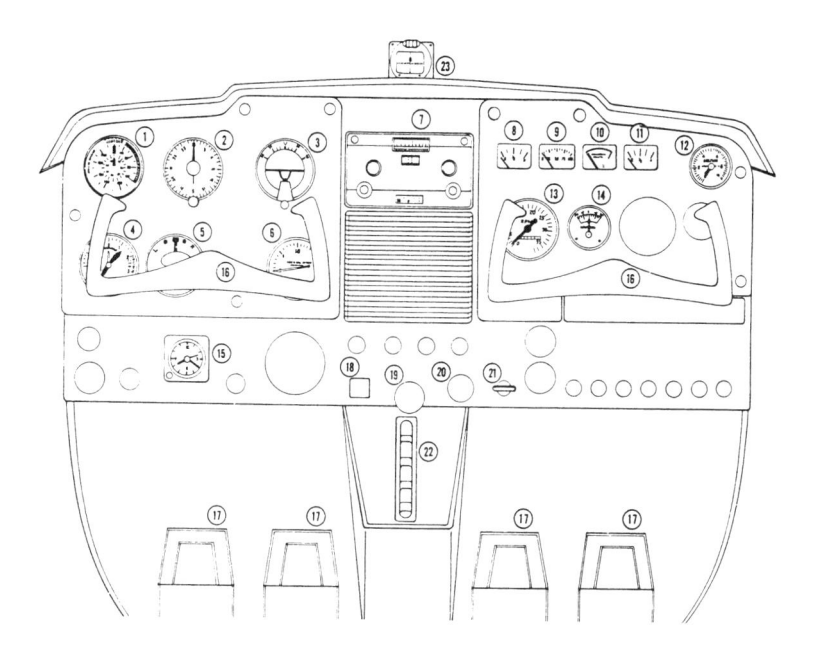

VFR Instrument Panel

1. Airspeed indicator
2. Gyroscopic compass
3. Artificial horizon
4. Altimeter
5. Turn and bank indicator
6. Vertical speed (rate-of-climb-descent) indicator
7. VHF navigation — communications radio
8. Fuel gauge (left tank)
9. Oil pressure gauge
10. Oil temperature gauge
11. Fuel gauge (right tank)
12. Suction indicator (run by vacuum pump, which activates gyroscopic instruments)
13. Tachometer (measures revolutions per minute of propeller)
14. Battery — generator indicator
15. Clock
16. Control wheel (dual)
17. Rudder pedals
18. Carburetor heat control
19. Throttle control
20. Fuel-air mixture control
21. Wing flaps control
22. Trim tab control
23. Magnetic compass

This panel represents a basic VFR panel with one navigation/communications radio.
(Drawing courtesy of the Department of Transportation)

pitot tube on the wing, giving the pilot raw data but not the bottom line. To get true air speed, temperature and altitude have to be factored in. Then, to get true groundspeed, wind direction and speed have to be known too. So the reading on the gauge won't tell you how soon you'll get to Grandma's unless you know what to do with this information.

As you become more familiar with flying, you might enjoy working with the information you see on the panel to see how close you can come to computing ground speed.

On the face of the air speed gauge, you'll also notice a white arc (for airplanes with flaps, this arc indicates the operating range in which flaps can be extended) plus a green arc and a red line. The red line indicates a speed not to be exceeded (i.e. the airplane's structure has not been rated to exceed this speed). Don't get nervous if the pilot gets within a hair's width of the red. It's an acceptable way to get a clean airplane down quickly in smooth air. The green arc is the normal operating range including the minimum speed required to stay airborne.

Rate of Climb

Although this appears to be reporting how many feet per minute the aircraft is climbing or descending, don't take it as gospel. The pilot doesn't fly by this instrument because it's a time lag reference that tells how many feet per minute the airplane has just gone up or down. Still, it's of interest as an indicator of climb performance, and it helps the pilot make the descent more gradual for sensitive ears. (ROC may also be labeled VVI, or Vertical Velocity Indicator.)

Compass

The old, basic magnetic compass is still required cockpit equipment, and you'll see it near the windshield. This compass has been adjusted (or "swung," a term you don't hear much in aviation) for this airplane, so you'll see a card attached listing correction factors for this particular installation.

The only time it reads correctly, however, is when you're sitting still or flying perfectly straight. Even the slightest turn sends it swirling, so the Right Seat may as well disregard it much of the time.

While we're on the subject of windshields, let's talk about the greenhouse effect. The glare can be horrendous and if you were in a car you'd just hang a beach towel over the offending window. In the cockpit, however, 360-degree visibility is crucial, and it takes precedence over sunshade placement. Get a good set of sunglasses. Don't obscure any of the pilot's window area. And, if you can't stand the heat, get out of the kitchen.

Gyro Instruments

Because they are stabilized by internal gyroscopes, these indicators remain constant despite the airplane's twists and turns. By using them, the pilot has a complete picture of what the airplane is doing even if you're flying through solid pea soup and can't see the ground, the horizon, or any other outside reference.

A note about pronunciation. If we're talking sandwiches, gyro rhymes with Nero. In aviation, gyro rhymes with Cairo.

Artificial Horizon

Passengers love this instrument because it shows a little airplane and a "horizon." You climb or turn and the little airplane appears to climb or turn. Under instrument conditions, passengers can still picture the airplane flying normally. So, if a passenger has motion sickness or vertigo and can't see a horizon or any other reference to the real world, staring at this instrument sometimes helps restore orientation.

The brain, aided by henchmen in the inner ear, can tell lies to the stomach. You can be right side up and flying on the level, yet your brain tells you quite convincingly you're upside down. If you've ever experienced vertigo, or a complete loss of spatial reference, you'll understand the problem if a passenger develops it. Hyperventilation, panic, and outright hysteria can result and no amount of comforting or logic will convince that passenger that you're not spiraling through space into oblivion.

Here's where the artificial horizon can save the day, so it's not a bad idea to call it to passengers' attention once in a while so they'll get used to using it as a rope they can grab onto if they feel they're sliding into some black hole. The real horizon is, of course, the surest reference. When that can't be seen, use the artificial one.

Gyro Compass

Because it is immune to the topsy-turvy problems of the magnetic compass mentioned above, the gyro compass is a more stable indicator. If it's a simple type, you'll see the pilot resetting it about every 15 minutes to adjust for precession of

The Instrument Panel

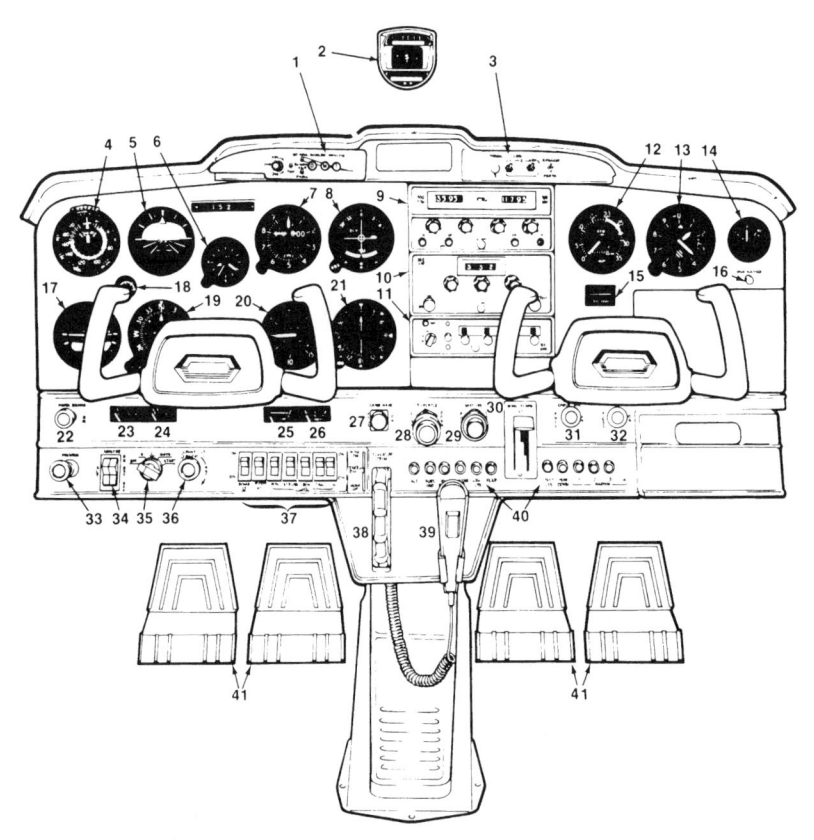

Instrument Panel

1. Marker Beacon Lights
2. Magnetic Compass
3. Audio Switching Panel
4. Airspeed Indicator
5. Attitude Indicator
6. Clock
7. Encoding Altimeter (Electric)
8. Omni Bearing Selector
9. NAV/COM (Radio)
10. Automatic Direction Finder (Radio)
11. Transponder
12. Tachometer
13. Altimeter (Barometric)
14. Battery/Alternator Indicator
15. Hour Meter
16. High Voltage Warning Light
17. Turn-and-Bank Indicator
18. Suction Indicator
19. Heading Indicator
20. Vertical Velocity Indicator
21. ADF Indicator
22. Parking Brakes
23. Fuel Gauge (Left Tank)
24. Fuel Gauge (Right Tank)
25. Oil Temperature Gauge
26. Oil Pressure Gauge
27. Carburetor Heat Control
28. Throttle Control
29. Fuel Air Mixture Control
30. Wing Flaps Control
31. Cabin Air Control
32. Cabin Heat Control
33. Engine Primer
34. Master Switch
35. Ignition Switch
36. Panel Light Control
37. Interior & Exterior Lights
38. Trim Tab Control
39. Microphone
40. Circuit Breakers
41. Rudder/Brake Pedals

This is a fairly representative diagram of an IFR instrument panel.

the gyro; more sophisticated types are slaved to a magnetic compass and are self-adjusting. You'll see a gyro slaving gauge on the panel. The gyro compass is sometimes called a Directional Gyro or DG, or a Heading Indicator.

Turn Indicator or Turn and Bank

Two types of indicators are used to indicate rate of turning. More modern types show a tail-on view of an airplane with little markers at the wing tips. Older style turn-and-bank types have a vertical needle and two little doghouses that, when matched up, give a standard turn rate of 3 degrees per second. Both types have a curved tube with a ball in it. It's simply a nonsensitive version of a carpenter's level, which tells the pilot whether the airplane is slipping or skidding.

Gyroscopes

As children, most of us had a toy gyro that looked like a high-tech version of a "top" — a toy that, once started, continues spinning on its own for a long time. In a gyroscope, a heavy wheel, mounted on an axis in such a way that it can turn freely in all directions, rotates very fast. It is able to maintain its own equilibrium on this axis, even when the mount itself is rising, falling, or tilting. Thanks to this balance, gyro instruments are able to operate unaffected by the motion of the airplane.

The gyro is subject to a phenomenon known as precession, a type of deviation that causes the mount to change position ever so slightly. In high school physics, you learned about precession by holding a spinning bicycle wheel while standing on a turntable. The axis of the spinning wheel wanted to stay in the same position; if you tried to move the axis, a secondary rotation was created.

To make up for this precession, unless the airplane has a slave compass as described in chapter 6, the pilot resets a gyro compass at intervals, usually about every 15 minutes.

Engine Instruments

All engine instruments are marked with red lines, green arcs, and sometimes a yellow arc. Their meaning (stop, go, and caution) is obvious, and most readings are easy for passengers to understand. Don't get too concerned when readings are in the yellow on a hot day. The pilot knows they'll settle down when enough air flow reaches the engine.

Every instrument panel has three gauges that are familiar to anyone who has a car: Oil Temperature Gauge; Oil Pressure Gauge; Tachometer (shows engine RPM).

The panel may also have:

Cylinder Head Temperature Gauge

It's comparable to the coolant temperature gauge in a car but, since the airplane's engine is air cooled, it measures the temperature of the cylinder head. If the airplane is liquid cooled, the gauge acts exactly like the same gauge in a car.

Exhaust Gas Temperature Gauge

The EGT helps the pilot set the fuel mixture correctly. Without such a gauge, the pilot goes by the sound of the engine and the seat of the pants. With a gauge, however, the pilot can go for the perfect compromise between too lean and too rich a mixture. Too cold a temperature and too rich a mixture makes soot. Too cold a temperature and too lean a mixture can burn the valves. Too hot an exhaust temperature can melt the pistons.

Ideally, the EGT will be about 25 degrees cooler than the maximum achievable temperature, guaranteeing that you're just slightly to the rich side of the optimum combustion range.

Ammeter or Voltage Meter

They perform the same function as your car's meters, and monitor the airplane's electrical system.

Vacuum Gauge or Suction Indicator

This shows the supply of vacuum being provided by a pump on the engine to run pneumatic gyro instruments. Output is controlled by a regulator and shouldn't change at all once the engine is above idle speed. If the pump fails, this reading goes to zero and the pilot has to rely on electrically powered instruments or an alternate source of vacuum. A higher-than-normal reading on this gauge usually means that the air filter is clogging. Too low a reading probably means there is a leak in the system or the pump is wearing out.

Hour Meter

This gauge shows how many hours the engine has run, to make it easier to anticipate routine maintenance needs, but it isn't used in navigation.

Fuel Gauge(s)

Usually there's a gauge for each tank, or sometimes one gauge for multiple tanks, with a switch showing which tank it's reading. Don't assume that the pilot uses this gauge the same way you use the gas gauge in a car. It's a handy reminder, but pilots compute fuel down to the last pound, by knowing what is being consumed at every stage of the flight.

A common source of anxiety for passengers is to see a fuel gauge reading Empty when in fact the pilot has already switched to another tank. A knowledgeable voice from the Right Seat can set them straight.

Cabin Heat Control

In most cases, there's never enough cabin heat in a small airplane. So, every ten minutes turn to face the back seat and repeat after us, "The knob is pulled out all the way."

The cabin heat in most light, single-engine airplanes comes from the exhaust manifold, using ram air. When you're sitting on the ground, very little air is coming through unless the heater has a blower, so everybody freezes. In climb, there's plenty of heat but little air flow because of the low air speed, so everybody freezes.

Then you reach cruise, and there's plenty of air flow but not as much heat, so everybody freezes. Letdown is worse still, with much too much air flow and too few BTU's. Everybody freezes.

Most light twins and some single engine airplanes use thermostatically controlled gasoline heaters, which can easily be managed by the pilot. However, in an airplane with a manifold heater, constant twiddling is needed to make the most of the heater's meager resources. It's a perfect assignment for the Right Seat, so the pilot may check you out on it. Right Seaters can also help keep a check on all the cold air vents (they probably look and work like the ones over every airliner seat) to make sure they're closed tightly when you're trying to keep warm.

Engine Controls

News flash. There's no gas pedal on the cockpit floor. In a car, you are constantly using the speed control but in an airplane a continuous throttle setting is chosen on the basis of a great many different factors.

You'll notice that everything on the panel is arranged so that a forward push increases speed, a pullback decreases speed. Push the yoke, go faster. Push the throttle, go faster. Push the mixture, the trim tabs, the propeller control, the carburetor heat. All of them affect speed.

In short, many controls in an airplane work backwards to the way they work in a car. Not to worry. It's all arranged so that, at full power, all the controls are firewalled. If the pilot forgets something, it sticks out like a sore thumb.

While we're on the subject of thumbs, we'll give another rule of thumb. Generally in American airplanes, a switch is up when it's On, and down for Off; knob controls usually increase to the right, decrease to the left.

Like all rules, these have exceptions, of course, especially in non-American airplanes. The Brits, who persist in driving on the wrong side of the road, get all their switches backasswards too. One legend says that the reason for this is that if the RAF were so badly shot up during a sortie that the airman didn't have strength to lift a switch, his hand could fall down upon it.

Throttle

Just like the accelerator on a car, it controls the amount of power the engine generates.

Mixture Control

All but the simplest airplanes have a control that determines how much fuel will be mixed with the air. This is necessary because, as elevation increases, atmospheric pressure is less. So, the higher you go, the less fuel is needed for optimum combustion.

This control can also be used to stop the engine. Pull it out, and it cuts off the fuel. That is why, on some control panels, the mixture control is labeled *Idle Cutoff.*

More and more small aircraft are now going to a throttle arrangement like that used in larger planes, in which levers are used instead of push-pull knobs. Uses and results are the same.

Propeller Control

If the airplane has a controllable or constant speed propeller, another knob or lever operates its governor, which increases RPM by decreasing pitch (i.e. the bite that the blade takes out of the air).

If you are used to a tachometer on a car, this can really buffalo you because, in a car, you call for more throttle and the RPM go up. In the airplane, however, the pilot can add beaucoup throttle with no change in the revs because RPM is controlled by the propeller governor.

To get more power in an airplane that has a constant speed propeller the pilot increases both the throttle and the RPM by using both the throttle lever and the propeller control.

Coming on the market now are multiple automatic systems that control overall power with a single lever, as is done on turboprops. If you encounter such a system, disregard all of the above.

Carburetor Heat

Even on a sweltering day, ice can form in the carburetor because of the cooling effect of the spray of gasoline. On a car, this is handled automatically, but in an airplane it's done manually because, when you want full power, you don't want some automatic device pulling the rug out from under you.

If the engine is fuel injected, carburetor heat loses its importance and is sometimes even labeled Alternate Air. In this case, it's used only if the air intake scoop on the nose becomes clogged with snow or ice.

Airplanes don't have chokes, which is one of the reasons why they are so hard to start when cold. In a car, a choke is used to choke off normal air flow to make a richer fuel mixture and thus easier starting. However, chokes can fail, and that means losing the engine just when you need it most. So airplanes use a primer, pumping in spurts of raw fuel as a starting aid, and all this is accompanied by coughing and sputtering that can convince passengers that the end of the world is at hand.

Another reason for hard starting is that aircraft engines are large, light, powerful and air cooled, which requires sloppier clearances than a car engine does. This in turn means heavier oil, and that means harder crankling when the engine is cold.

Hard starting goes with the territory, and it does not portend, as it often does in most other vehicles, that an engine is one step this side of the scrap heap. Once warmed up and running smoothly, an aircraft engine is one of the world's most reliable machines.

About APUs

Because batteries are heavy — they contain lead — aircraft are equipped with the smallest batteries that will do the job. And the job is not an easy one because aircraft engines are, for the reason outlined above, hard to start. Often available at airports, even small ones, is an outside power source known as an Auxiliary Power Unit, or APU.

If the pilot brings an APU into the start-up picture, don't think of it as comparable to using jumper cables to pull a car back from the jaws of death. It simply means that a perfectly healthy airplane with a perfectly healthy battery is being given a boost from a more efficient, less costly source of juice. In extremely cold weather, the pilot may also call for an engine heater.

Using these outside sources is a great way to extend battery life and reduce engine wear.

Electronics

The other controls on the panel include magneto switches, battery master switch, the alternator or generator switch, starter button(s), and a separate switch for each system in the airplane (running lights, strobe lights, landing lights, etc.). Everything is clearly labeled, and presents no mysteries to the Right Seat.

As we mentioned earlier, airplane systems don't just whir into action as they do in a car when the key is turned. Every action is planned and deliberate, leaving no room for error. That's why there are so many switches. Everything must be turned on and turned off.

Another panel, which isn't very interesting to look at and which doesn't appear to get a lot of attention from the pilot, has circuit breakers or fuses for each of these systems. This area is pretty much ignored unless there is an electrical problem or, in the rare cases where a circuit breaker is also used as a switch.

If you do see a popped circuit breaker, call it to the pilot's attention but never try to reset it on your own. The pilot may have popped it intentionally, or it could have opened because of an electrical problem that needs to be addressed.

Chapter 7

What's Going On Here?

We all learned in junior high that airplanes fly because of a victory of thrust and lift over drag and gravity. When the airplane is going fast enough to provide sufficient relative wind to the wings, it leaves the ground.

While it's tempting for us to dwell on Bernoulli's Theory of Induced Lift and Newton's Third Law of Physics, our guess is that you're more interested in what is really going on between the pilot and the airplane, the pilot and the tower, and the flight and you. So let's cut to the chase.

The Airport Scene

Most of us can jump into a car, start the engine as we glance at the gas gauge, and back out of the driveway with barely a backward glance. So it sometimes mystifies Right Seaters when the pilot wants to get to the airport an hour before flight time and then piddle around forever with what seems like pointless chitchat and endless checking.

The routine is more complex than it appears. It probably goes something like this:

Before Leaving Home

The pilot will telephone a weather office for a briefing and perhaps to file a flight plan. The itinerary has already been worked out, charts assembled, fuel needs figured and en route time plotted.

With luck, you or the pilot will be able to learn the price of fuel at your destination (it may pay to phone around), so a decision can made on how much fuel, within safety limits, will be taken on at each stop. The Right Seater can also phone around for motel and rental car reservations. If the trip will be over water, you'll need to round up life jackets and a perhaps a life raft. Provisions and other survival gear will be assembled, depending on the terrain and the type of trip.

What a Flight Plan Is, and Is Not

Although a flight plan must be filed for every IFR flight and for any VFR flight through an ADIZ (Aircraft Defense Identification Zone), pilots are encouraged to file one for every cross-country. It's done in writing if the field is a Flight Service Station; in most cases it's done by phone. It costs nothing (although there may be a charge for search costs if the pilot forgets to close it at the end of the flight)

Flight plans are among aviation's best free services, assuring that someone will be out beating the bushes to find you — and knowing which bushes to beat — if you don't show up on time.

By reviewing the flight plan, rescuers know:
- Your path of flight including time and place of departure, expected time of arrival, and your cruising speed.
- Time the fuel expired
- Type and color of the airplane
- How many souls are aboard
- The pilot's name, address, telephone number, and the home base of the airplane.

On longer flights, the pilot will probably check in at each FSS (Flight Service Station) you pass. So, if a search must be begun, officials have a pretty good idea of where to start looking.

We can't overemphasize the importance of remembering to close a flight plan. Forgetting it can set off a chain of events that is expensive to the taxpayer and terrifying to families back home. If you contribute nothing more to the flight than keeping a pilot from forgetting to close a flight plan, you have paid your passage.

Arriving at the Airport

Now that you've arrived at the airport, the most dangerous part of the trip is over. If the pilot hasn't already charted a route at home and filed a flight plan, it's done now. For a VFR (Visual Flight Rules, i.e. fair weather, in which you can see the ground) flight, the charts used will be either sectionals or WAC charts. In an instrument operation, even for a VFR flight, low altitude En Route Charts are simplest to use.

If the pilot describes to you the route from, say, DeLand, FL to Kinston, NC, it will sound something like, "Direct Ormond Beach, Victor 437 to Stary Intersection, then Victor One to ISO." By the time you finish this book and have taken a few flights, this kind of lingo will begin to make sense.

If the right and left seat go over the flight route before takeoff, it will be easier for the non-pilot to follow along once aloft.
(Photo courtesy of Jane Harding, Bengtsson, Inc.)

If a sectional or WAC chart is being used, the pilot will use a plastic plotter to compute the bearing on the compass rose around each Omni station, then measure the distance along the course. This provides the information needed to compute time and fuel. It's easier with En Route charts because bearings, distances, and safe altitudes are already listed.

Next, the pilot goes to the airplane to do a brief walk-around inspection and to check fuel. Don't assume that all of you can drive right up to the airplane to off-load your gear. At some fields, security rules keep all cars off the ramp. At others, a ramp pass can be issued — often in addition to a key or transmitter for opening the gate.

All this takes time, but it's only the beginning of what appears, to the infrequent Right Seater, to take longer than gum surgery. Now the pilot will check to see how much fuel is aboard. Even if it's your own airplane, it's not safe to assume that there is as much fuel aboard now as when you last left the plane. Evaporation accounts for some loss; theft or leakage could have occurred.

Before anything else is jostled, the pilot will drain the fuel sumps — just one more reason why nobody smokes around air-

The Pilot will check the fuel supply by actual measure, rather than relying on gauges.
(Photos courtesy of Embry-Riddle Aeronautical University).

All fluids are checked. If irregularities are revealed, they are discussed with a mechanic.
(Photo Courtesy of Embry-Riddle Aeronautical University)

planes that are getting ready to launch. This is a critical step because any water or crud in the fuel will have settled to the sump, a section at the lowest point of the fuel system, where sediment collects. This fuel must be drained off and examined.

Few pilots will relegate this task to anyone else; bad fuel means a sputtering mill, always at the worst possible time. So, most pilots scrutinize the sump squeezings scrupulously. They'll be looking for water, metal filings, and impurities of any kind that could point to impending engine problems in the engine or fuel itself.

It's time now to order fuel, which can be done by calling for the fuel truck or by taxiing to the fuel pump. Oil and other fluids will be checked and ordered at the same time. Because the pilot is responsible for the amount and grade of fuel being put aboard, and which tanks are being filled, it's not coffee break time yet. The pilot will want to stick with the airplane because each type of avgas is tinted a different color, so the pilot makes triple sure it's the right octane by reading the truck, reading the hose, and noting the gas color. After seeing that the right amount of fuel is put into each tank, the pilot checks personally to see that the caps are secured properly.

62 What's Going On Here?

Even this is a job that won't be relegated to anyone else. In a car, a loose cap would simply fall off along the highway, a little fuel would slosh out, and at the next stop you'd plug the hole with a corncob. However in most airplanes, an improperly fitted cap can allow fuel to siphon out because low air pressure on the top of the wing creates suction. Worse still, bladder-type tanks rise as the fuel is ejected, and continue to read Full. Your first clue that you're out of gas is when the airplane suddenly turns into a glider.

Lastly, the pilot may or may not drain the sumps again. Now it's time to go back into the office to pay for the fuel, have a cup of coffee, get a weather update, use the john, and file a flight plan if it hasn't been done yet. If it's a small, unpaved field, locals may also be able to share some tips about the condition of the sod.

The Walk Around Check

It's SOP (flight jargon for Standard Operating Procedure), on every airplane from puddle jumpers to jumbo jets, for the pilot(s) to walk completely around the aircraft looking for many specific things as well as for any abnormalities.

"I often take a passenger on the walk around," says Bruce

During the preflight, tires are checked for inflation and damage.
(Photos Courtesy of Embry-Riddle Aeronautical University)

The Cockpit Companion

As part of a preflight check, these men are looking over the stabilator, a combination stabilizer/elevator found on a few airplanes.
(Photo Courtesy of Embry-Riddle Aeronautical University)

Checking to see that the elevator moves its full travel.
(Photo courtesy of Embry-Riddle Aeronautical University)

During the walk-around check, every moveable part will be examined. (Photo courtesy of Embry-Riddle Aeronautical University)

Jenner, who flies his MU-2 with his family and friends aboard. "Sometimes a first-timer gets nervous with a pilot who doesn't have epaulettes and a hat. So I explain everything."

First, all control locks, sun shades, and other covers are removed and stowed; the Right Seater might be asked to help remove and stow tiedown ropes and chocks.

During this inspection, the pilot will look at the engine, checking oil and looking around under the cowling for any loose gear, bird nests, or oil leakage. If the tires are soft, they are filled by using a hand pump or an air bottle. The pitot tube (a tiny air intake) is checked. Mud daubers may have established a base camp inside since your last flight; in winter it may be clogged with ice. The prop is checked for nicks or other damage, then the pilot looks at the leading edges and tips of the wings including nav lights.

Your help in checking the lights will be appreciated. As switches are flicked in the cockpit, you can verify that all the lights, strobes, and beacons are operating. It's quicker, and far less battery drain, if two people do this together instead of one person operating switches and then running around to see what is lit.

On the aft side of the wings, the pilot will check control surfaces and the condition of the skin. All moveable parts will be checked around the fuselage and tail. If any control locks (braces that are used to disable moveable surfaces while the aircraft is stowed) have been overlooked, they'll be found and

The Cockpit Companion

Don't attempt to close the door unless you've been checked out on it's operation. It's not like slamming a car door.
(Photo courtesy of Embry-Riddle Aeronautical University)

removed now. Usually, control locks have tags or flags to make it almost impossible to overlook them.

The basics are the same; aircraft differ. The more complex the airplane, the more areas the pilot will want to poke, scrutinize, tweak, uncover, and contemplate.

The pilot will probably choose to stow all the luggage personally; he wants to know what each piece weighs and where it's stowed to help in weight and balance calculations. He might also assign seats according to passenger weight, for balance purposes — a touchy process at times if Binky wants the seat with the most legroom but Ralph the Rhino has to be put there to keep the center of gravity from moving 4 feet aft of the fuselage.

By now, you're well aware that flying Right Seat in a private plane is different from airline flying. All these same preflight steps take place on the airlines, but you don't see them happen because you're in the terminal calling your answering service and having one last cup of coffee. When it takes place on the tarmac in blast furnace heat, or on an icy day, it seems to take forever. Still, the pilot will see it through from start to finish, every time.

In the Cockpit

You're now ready to board the airplane, in the order directed by the pilot. This is a good time to note the placement of handholds and No Step areas, with a reminder that seat backs should never be used as grips. A passenger, using the back of the Left Seat to hoist himself forward to ask a question, will not only scare the stuffing out of a busy pilot, he could break the seat. The seat back wasn't designed to be used as a monkey bar.

The order in which you all board the airplane usually is unrelated to seniority, courtesy, or chain of command. It may be dictated by the seating layout, or by the weather. On a blistering day, the pilot may elect to board first and get some systems operating before stuffing the passengers aboard. In any case, when it's time to close the door, the pilot will do so. So don't try to be helpful unless you're well checked out. It's not like slamming a car door. Aircraft doors usually need some special body english to secure all latches and locks.

"I do make sure that passengers know how to open the doors," adds Bruce Jenner. Even though the pilot will probably ask that the door stayed closed until an "all clear" is sounded after the flight, it's always wise to make sure all able bodies know how to get out of the plane in a hurry if necessary.

But let's get back to the start-up routine. All aircraft require certain checklists, and may involved separate lists for pre-start, pre-taxi, pre-takeoff, cruise, climb, descent, landing, post shutdown. The simplest airplanes, those without electrical systems, usually have just a magneto switch and a fuel valve, so there is little chance to go astray.

As airplanes get more complex, however, the checklist gets longer and longer. In any case, the routine is far different from anything you've experienced in a car. You get into a car, turn the key, start the engine, and everything works. In an airplane, everything has to be turned off or on separately.

For example, an airplane with an electrical system will have a switch for the generator or alternator, which has to be turned on to generate electrical power. Then there are switches that deliver this juice to each separate item: boost pumps, hydraulic pumps (which, in some cases, have to be on before the parking brake can be set), fuel valves to the tank that will be used during engine start-up, and so on.

About Checklists

The name implies a list on which check marks are to be made as each listing is verified. If a checklist is actually used this way, it's difficult to overlook an item. But wait! The checklist is probably not like a grocery list that can be ticked off, but is a permanent, laminated card that the pilot goes through with fickle finger, item by item. It's easy to miss one, which is why the Left Seat probably won't ask a non-pilot in the Right Seat to read the list.

This is just another area in which it's lonely at the top. Don't try to help unless, perhaps, the panel has a mechanical checklist in which a toggle is set as each item is checked. Used in reverse, the same mechanical checklist serves as a pre-landing check. With a clearly visual check, both Left and Right Seat can see that this type list was completed.

In any case, this is a time not to divert the pilot's attention. Running through a checklist becomes a religious routine that

Many pilots make up their own acronyms for checklists. This one, based on the word CIGARTIPS, is available commercially.
(Photo courtesy of R&M Affiliates)

involves rote, habit, memory, and logic, so it should be accomplished in a seamless, uninterrupted series. Even in the most simple airplane, in which someone is propping for you, the pilot has to be fully attuned to what is happening at the prop. Here's where the Right Seater can help. As the pilot shouts, "Clear the propeller(s)!", lean well to your side to see if anyone or anything is in way. On a self-starting airplane too, the pilot has to be sure that no one is in the arc of the propeller, so an extra pair of eyes is appreciated.

The next step is to get the fuel mixture right, get the engine firing, and keep it firing. Airplanes are, by nature, difficult to start for reasons we explained in Chapter 6. People who are hearing this belching, smoking, coughing, sputtering mayhem for the first time are convinced that they're about to meet their doom.

Just keep telling yourself, and any nervous passengers, that the Spirit of St. Louis sounded a lot worse than this, and she made it all the way across the Atlantic. Once things settle down, aircraft engines are about the most reliable machines going.

The first few seconds of engine start are critical. As soon as the engine catches, the pilot's eyes go to the instruments to get immediate feedback about whether pressures and temperatures are coming up correctly. If you have been checked out ahead of time on what the pilot is looking for in these indicators, you can serve as another set of eyes on the instruments while the pilot gets a clearance or talks to ground control.

CIGARTIPS

Every aircraft must have its own checklist, based on the model, powerplant, equipment, and accessories, so it would be impossible, and inappropriate, to give here an explanation of everything that could appear on every checklist. However, we can explain the simple CIGARTIPS checklist shown, which includes the checks you're most likely to encounter.

Keep in mind, though, that these are just general categories. Every pilot will have many personal categories, sublists, prejudices, practices, and perhaps a rabbit's foot.

Controls You'll remember that, during the walk-around check, the pilot examined control surfaces for free movement. Now they're checked in the cockpit by working the yoke and rudder pedals and observing the movement of the surfaces.

Instruments Other instruments will come later, under Runup. At this point, the pilot will deal with instruments that have to be set, such as the altimeter and the gyro compass.

Gas Make sure the gas gauge agrees with the amount of fuel that was put aboard, and that the selector valves are on the right tanks.

Attitude The pilot will set the elevator and trim tabs to accommodate the weight and balance of the airplane for the current load. In most small airplanes, the pilot does this by the seat of the pants. In heavier airplanes, settings may be prescribed by the manual for a given weight and balance.

Runup The pilot will check now to see that the oil pressure, cylinder head and oil temperatures, and vacuum gauges are reading correctly. Also checked at this time are the alternator (ammeter/voltmeter readings correct?), magnetos (both operating and firing all spark plugs), propeller (if it's variable pitch or constant speed, it will be run through a pitch change to make sure it's working right and that all cold oil has been flushed out of the hub) and carburetor heat (it's applied to find out if any carburetor ice is present and, if so, to melt it. Then it's checked to make sure it's off for takeoff.)

Traffic One of the most important references for the pilot, as well as for the Right Seater who wants to help, is a check of traffic that is on, approaching, and leaving the airport, giving the pilot a complete mental picture of the pattern.

Information Each pilot will have a personal list of information to be reviewed at checklist time, such as local weather, weather in route, weather at the destination, and so on.

Performance During the checklist, the pilot may take this one last opportunity to compute again whether the runaway is suitable for a successful takeoff with this load and at these temperatures.

Safety One last check is made to see that all doors and windows are closed, and seat belts securely fastened.

Many pilots have their own acronyms, probably many of them unprintable — we remember one called UP YOURS —developed for a particular airplane. One popular, fun-to-say acronym is the GUMPSAC checklist, which generally (but not always for every pilot or every airplane) stands for:

Gas tank (see above)

Undercarriage (do indicators show that the gear is down and locked?)

Mixture (an appropriate carburetor setting will be chosen)

Propeller (see CIGARTIPS, above)

Seat belts

Autopilot (the pilot will want it disconnected for takeoff and landing)

Cowl flaps (They control air flow to the engine, so the pilot will want them full open for takeoff.)

Taxi and Takeoff

Some airplanes have radio speakers from which there is no escape. They are usually too loud and too garbled, and they blare forth at odd intervals, scaring the bowels out of everybody.

If earphones are available and you have a choice, request a headset. You can now hear what's going on between the pilot and whoever is on the other end of the conversation, and you can adjust the volume to suit your own hearing.

A new wrinkle, introduced by PCR Systems (6851 Hwy. 73, Evergreen CO 80439) is a recording device that makes a superb tool for the pilot and a useful toy for the Right Seat. Put into the mike wire in place of the PTT (push to talk) switch, the PTR

The Cockpit Companion

With the new push to record (PTR) switch, which is a direct replacement for the push-to-talk (PTT) switch found in most aircraft, the pilot can record 5 minutes of radio transmissions. It's invaluable for playing back and verifying fast-paced transmissions.
(Photo courtesy of Colorado Computer Associates)

[TM] Switch instantly records up to five minutes or 15 messages.

For the pilot who missed parts of any transmission, the switch allows instant replay as many times as necessary. If any live messages come in while PTR is playing, they are muted but still audible. Or, in a stereo headset, messages come into one ear only.

The PTR Switch has other features and is available in several models. The value to the Right Seat is that, once the pilot gets the message, you can continue to play back the recording until it begins to make sense to you. Most radio transmissions come across as garbled gibberish to the unpracticed ear. In time, you'll begin to make sense out of them.

At a small airport, the radio is now tuned to Unicom (the simple, local airport communications radio) or nothing at all. The pilot taxis to the end of the runway, in a maneuver that will probably surprise you because nobody's hands are on the yoke or stick. Steering on the ground, in almost all airplanes, is done with the feet on the rudder pedals/brakes.

Now the engine is usually run up fast enough to assure that all systems, including both ignition systems, are operating. Generators/alternators are checked for output, carburetor heat or alternate air are checked, and the pilot watches to see that temperatures and pressures are in the green.

Trim tabs and flaps are set. Freedom of travel of all controls is checked once more. The pilot then glances around to make sure everyone is wearing a seat belt and the door and windows are secured.

While all this is going on in the Left Seat, the Right Seat can scan the area for traffic. Two heads are better than one. If you do see something, simply point, just in case the pilot missed it. According to Bruce Jenner, he turns it into a game for his family. "I always bet them that, if there is an airplane in the area, I'll spot it first."

Just before starting the takeoff run, the pilot will probably make a blind announcement over the radio that the airplane is about to take off in so-and-so direction on so-and-so runway. This gives the radio audience, on the ground and aloft, a chance to speak up in case there's a conflict.

In a car, the trip begins slowly but in an airplane, full throttle is needed to get off the ground. All this is accompanied by horrendous noise, rattling, shuddering, buffeting, and other pandemonium that may be frightening to passengers who are new to small plane flying. Hang in there. As soon as you're airborne, the noise eases.

It's now that the most critical part of the flight — reaching a safe cruising altitude — begins. The pilot is concentrating to the fullest on engine sounds, instruments, traffic, retracting the gear at the right moment, and keeping one eye open for a place to

In a car the getaway is gradual but in an airplane the engine is thoroughly warmed up before takeoff.
(Photo courtesy of the Beech Aircraft Corporation)

put down if the engine quits.

On some airplanes, retracting the gear sets off a series of lights that sends passengers the wrong signal. We once had an airplane in which a red light came on as soon as the gear was up and stowed. The intent was to say to the pilot, "Don't land, you bozo! The gear is up!" A green light indicated that the gear was down, i.e., "It's OK to land now."

To passengers, however, green meant go/good news and a red light said, Stop! All hell is about to break loose! Here's where a Right Seater can anticipate any passenger anxieties and explain things calmly.

Anything short of a mushroom cloud in the direct path of flight should not be brought to the pilot's attention at this time. In a normal transition to climb power, the pilot will usually throttle back and/or decrease the RPM (with a controllable pitch propeller), so you can expect some changes in sound and on the tachometer.

Once a safe altitude (i.e. clear of TV towers, skyscrapers, mountains and other obstructions) has been reached, the flight settles down. The climb will be completed and the transition made to cruising.

When you hear the engine settle into a smooth purr and see the pilot relax a little, you know the rest will be fairly routine. At this point, the passive Right Seater may want to open a book and veg out. The more involved Right Seater may want to stay alert, look, listen, and learn. For example, it's a help if you help manage the charts.

The Flight

Cruising altitude depends on several factors, some of them the pilot's choice and others dictated by traffic, regulations, clouds, or terrain. You may think the airplane should be at less bumpy altitude, or one with a better view, but it isn't that simple.

Flying VFR at more than 3000 feet above ground level (AGL), eastbound flights should be at odd thousands plus 500 feet; westbound on a course between 180 and 359 degrees, the pilot flies at even thousands plus 500 feet. IFR traffic is flying at assigned altitudes at odd thousands eastbound and even thousands westbound. Got that? Actually, you don't have to remember it all. We just thought you'd like to know that everyone isn't just ramming around up there, ad lib.

In addition to complying with these rules, the pilot has chosen a particular altitude with consideration to headwinds, tailwinds, crosswinds, cloud tops, cloud bottoms, freezing level, turbulence, obstructions, and precipitation.

With these factors in mind and an altitude chosen, the pilot now can level off and set the engine power either for maximum speed, maximum fuel economy, or any point in between, depending on conditions at hand.

Engine temperatures will now stabilize and remain constant, but fuel gauge(s) will change. The pilot may ask you to speak up if you see a gauge reach a certain point. In any case, don't panic if a gauge reads *Empty*. It's normal to run a tank completely dry before switching to another. This allows the pilot to calculate exactly the fuel remaining.

The Left Seat's tasks now include monitoring the flight, navigation, keeping abreast of the weather, and any radio communications.

Navigation: Needle in a Haystack

The charts are totally incomprehensible and you may not be able to pick out any identifiable railroad tracks or interstates on the ground. How does the pilot know where you are?

By following along on a road map, you may be able to keep up with the route, town by town, lake by lake. However, things will be clearer if you follow along on an aviation chart — a sectional, which is 1:500,000 scale; a WAC chart, which is 1:1,000,000 scale; or one of the terminal charts covering a small area, e.g. metropolitan Chicago.

Such charts provide a wide range of information that affect the flight — restricted areas that the airplane may not overfly (usually military), elevations and obstructions, major visual landmarks such as lakes, rivers, railroads, highways, major power lines, shorelines, towns and airports; instrument airways that can be flown using electronic aids; and latitudes and longitudes.

If you're interested in navigation and learn to follow a chart, you can tell exactly where you are at any given time. It's interesting to the casual sightseer, and it also helps the Right Seat answer questions from the Back Seat. By acquainting yourself with these charts you'll see why, if there is any engine trouble, the pilot knows immediately where to find the nearest airport by using both visual clues and electronic aids.

Charts

The most common aeronautical charts (don't say maps) used in smaller, slower airplanes are sectionals, covering large sections of the country. Most of New England, for example, is shown on the New York sectional; most of California is covered in two sectionals, San Francisco and Los Angeles. Charts are ordered by name, e.g. Jacksonville or Cleveland, and are done on a scale of 1:500,000. They contain topographic information, visual checkpoints, navaids and radio frequencies, landing fields, and controlled and restricted areas.

WAC (World Aeronautical Chart) charts, which are handier in faster airplanes, contain much the same information used in sectionals but on a scale of l:1,000,000. In other words, details are smaller but you need fewer charts to fly long distances.

VFR Terminal Area Charts are used by pilots of both slow and moderate speed aircraft for visual navigation in terminal control areas. Their scale is l:250,000.

En route Charts come in both low altitude and high altitude versions. They are used for instrument flying above or below 18,000 feet and are revised very often.

Other charts you may hear about are **SIDs** (standard instrument departures), **STARs** (standard terminal arrival routes), **jet navigation route charts**, and **instrument approach plates**, which are published separately for each approach at each airport.

> Promotional charts, which are not legal for navigation but are very handy references when you're trying to land near a particular attraction, are often published by tourism agencies or resorts that solicit fly-in business. Road maps are also of interest to passengers because they are familiar aids on which non-pilots can pick out major cities, highways and waterways.

The Electronic Revolution

Even small airplanes on modest budgets today have space-age electronics on board. Here's just a sampling of the technology the pilot may be using:

ADF

Every IFR equipped airplane must have at least one Automatic Direction Finder. It's an old system perfected before World War II, simple and reliable, that uses a low/medium frequency receiver incorporating a direction sensitive antenna so that, when a station is tuned up, a small pointer superimposed over a compass rose points to the station selected.

ADF's real value to the VFR pilot, especially when flying remote areas such as the West Indies, is that it can also be set to point to any commercial AM broadcast station in the world. Most of the time, the pilot is not using this radio so, if you're following along on a chart and if you have the pilot's permission, you can use the ADF to listen to the ball game, pick out stations on the map, take bearings, and generally entertain yourself.

ADF

> Radio signals are in the air all around, above, and below you. Some are there for the sole purpose of guiding aircraft; others are broadcasting the Boston Pops or are used by lake or ocean shipping.
>
> The ADF (Automatic Direction Finder) uses a directional antenna that locks on to the station the pilot has

tuned up, if it's within range. The antenna is connected to an indicator on the instrument panel that looks like a compass rose with a needle in it, and shows the relative bearing to the station.

If you were flying north, the needle would give the magnetic bearing to the station. So, if you're flying any direction other than north, the pilot will either compute the bearings in his head or, if the instrument has a rotatable face, will turn the compass rose to the airplane heading and the heading under the needle will show the bearing to the station.

Conversely, the back end of the needle shows the bearing from the station. The ADF is used to find a station at a locale you're headed for, or to indicate a station that serves as a fix you want to pass abeam. Because it's usually the only set in the airplane that receives broadcast bands, it can also serve as passenger entertainment or to find islands or other remote spots where AM radio is the strongest signal available.

Sophisticated indicators are available that tie the compass rose to the slaved gyro compass so that it always indicates the direction of the airplane. Therefore, the ADF needle would always show the bearing to the station.

The pilot can use the ADF to home in on a signal, to fly directly away from a signal, or to take bearings from signals to triangulate a position.

VOR

The Visual Omni Range is the primary worldwide nav system for instrument flying. It uses a VHF (Very High Frequency) receiver, which means it is not subject to atmospheric interference as low frequencies are. It can receive only line-of-sight signals, so the higher you are the farther it can be used.

Imagine that the radio station is a railroad roundhouse with 360 tracks, one for each degree of the compass rose. The pilot dials up the frequency of the "roundhouse", listens to the code identifier to verify that is the right station, and sets in the desired course, or "track." By flying in relation to what the needle indicates, the pilot can follow a pre-selected course to the next waypoint or can dial up any station within range to get a bearing fix. Using two stations, a position fix can be made.

VOR, also called Omni

On the panel, you'll see a round dial with a compass rose and a vertical needle and, inside, a smaller meter or window with the words To and From. There may also be a couple of flags that drop down if the VOR is receiving no signal, or an unreliable one.

Here's how it works in slow motion. Picture a lighthouse that makes one sweep every six minutes. When the beacon passes through magnetic north, a light on the top of the lighthouse flashes.

To know what bearing you're on relative to the lighthouse, you'd start a stopwatch when the light on the top flashes. Then, when the main beam hits you, you stop the stopwatch and read how many seconds passed. This tells, in degrees, your bearing from the lighthouse.

All this is done for you, in milliseconds, by electronics, giving the pilot a constant line that can be flown on or near, or used as a cross-bearing.

Keep in mind that this signal is line-of-sight only so, when one station disappears behind a hill or over the horizon, the pilot will simply select another.

DME

Distance Measuring Equipment tells the pilot the distance to the station he or she has selected. A signal is sent from the transmitter in the airplane to a ground station, where it is received, decoded and identified. The station then uses its transmitter to send a signal back. The receiver in the airplane translates the time lapse information into nautical miles, which are shown on a display.

The pilot must do some interpretation, according to where

the DME is located. It's shown on charts, so you might look for it. For example, if the DME transmitter is co-located with the ILS (Instrument Landing System), which is at the far end of the runway, it will give the impression that you're still two miles away when you land on a two-mile-long runway.

Most DME's also indicate ground speed. DME ground speed is accurate, however, only when the plane is going directly to or from the station, so take this reading with a grain of salt unless you're fully acquainted with how it works. Angular distance has to be factored in. The closer you are to the station, the more it under-reports your actual speed.

Other Electronic Navigation

Loran, GPS, etc. are becoming increasingly affordable and common in general aviation. There are hundreds of different types and capabilities, ranging from the simplest units that report only latitude and longitude, to smart models that need only to be told where you want to go. They figure out everything down to the last ounce of fuel and foot of distance.

Many pilots, though, still make do with a simple circular calculator. When the pilot tells you, down to the last minute and mile, how the flight will proceed, don't be surprised if it's bang-on. It's simple math, in the hands of a skilled aviator who is using time, speed and distance, in concert with other factors such as wind speed and direction and outside temperature.

GPS

Loran and other satellite-based systems are being supplanted by the best, most accurate, and most widely available navigation system yet. It's GPS, for Global Positioning System. Relying on 21 satellites that will eventually be in orbit almost 11,000 miles above the earth, with between four and seven of them "visible" to your airplane at any one moment, GPS can pinpoint your position within an accuracy of about 30 feet, feet (in military modes, accurate down to a centimeter,) 24 hours a day, regardless of weather or interferences.

Its applications are exploding, not just in aviation but in boating and on land. Already in use are systems in which truck fleets are being tracked by GPS. Using this system, airports can even keep track of every baggage cart and catering truck on the field.

Basically, GPS determines your position by sending signals to satellites, measuring the transmission time by using clocks that are accurate to one nanosecond. By triangulating four ways, GPS reports your latitude, longitude, and altitude. (The fourth dimension is time, which we'll explain later.)

Remember in 8th grade, when we had to compute the height of the Washington Monument by using triangulation? GPS works in a high-tech version of the same problem, using radio waves that travel 186,000 miles per second to measure the distances to satellites thousands of miles away.

Your unit knew exactly what time the signal left the airplane. The problem was to know exactly what time each signal left the satellite. So, both the satellites and your receiver generate a complicated set of what are called pseudo random codes that can be interpreted by your receiver. The satellites, with their atomic clocks, are far more accurate than even the best clock in the average, consumer model GPS receiver, so the fourth satellite reading is used to eliminate the timing offset and put the other three readings in sync with your receiver.

To further assure accuracy the satellites, which orbit every 12 hours and are programmed with an almanac to keep them in sync with the heavens, are monitored twice a day by the Department of Defense. They are also fine-tuned regularly for any ephemeris errors, such as a gravitational tug from the moon. This constantly changing state-of-the-satellite data message is also sent back to your receiver, which sorts out this data as well as the pseudo random code information.

A great many other astronomical and mathematical principles are involved, such as atmospheric and ionospheric delays and a multipath error in which signals don't take a direct path back to your receiver. For now, though, it's probably enough to say that GPS is the best electronic aid yet to a sharp pilot who uses GPS and all other available aids in concert with sharply tuned horse sense.

Loran is an electronic, long-range navigation system that uses fixed radio stations to transmit, very accurately, timed signals in combinations, in such a way that a receiver can calculate its position by reading the time lapse between the signals and triangulating among various stations.

It all began during World War II. Unlike other systems, which are line-of-sight, Loran's very low frequency waves follow the curve of the earth, providing a range of about 1200 miles per station. At first, it was available only on both coasts for use by ships, but it became so popular with airplanes that blank spots in the center of the country were filled in with additional coverage. The original Loran-A was phased out when the improved Loran-C came along.

Basically, it tunes to whatever stations it can find, and tells the pilot where you are in terms of latitude and longitude, i.e. in degrees, minutes, and seconds. The pilot then has to translate this further, e.g. five miles east of Cincinnati.

Newer sets work with airport identifiers, storing their latitude and longitude in memory. So the pilot has only to dial in, say, LAX for Los Angeles, and the Loran tells you how far it will be, what compass course to fly corrected for variation for each area you'll fly through, and perhaps will also tell you, en route, where you are in relation to a rhumb line (a navigational line that cuts through all meridians at the same angle).

Some Lorans also interface with autopilots and do the entire job right down to the steering.

If you have had Pinch Hitter training, and can actually fly the airplane in an emergency, you may be taught how to use an emergency button on the Loran. Press it, and it will tell you the bearing and distance to the nearest airport(s) with a runway of 3000 feet or more. If you reject the first choice, it will give you another and another.

Even though the readout may appear to report latitude or longitude down to a tenth of a second (one tenth of a nautical mile), take it with a grain of salt. Loran wasn't designed to change signals as fast as airplanes can fly so, if it's within a half mile, you're doing well. The pilot who is flying on instruments is using Loran only in conjunction with more accurate, FAA-approved, radio navigation.

Many lightplanes now are equipped with most of the goodies airliners have.
(Photo courtesy of Beech Aircraft)

Radar

Because radar can see farther than the human eye, and through things that the eye cannot, it's an aid that can be useful to the pilot and is indispensable to the ground sta-

The Cockpit Companion 83

tions that are keeping track of your airplane, other traffic, and the weather. No FCC license is required to operate radar, but the sets themselves are licensed. Any work on them must be done by a licensed technician.

Most small airplanes do not have radar aboard. If yours does not, don't think you're in increased danger. Light plane pilots have so many other references to depend on, radar is by no means a must.

If you have a radar aboard, it consists of a transmitter, which sends out radio waves in pulses that travel at the speed of light and at the rate of 600-4000 per second, an oscillating antenna in a teardrop-shaped enclosure in the nose or under a wing to send out the pulses and receive the returning echoes, a receiver to collect and amplify the echoes, and the display, which is the miniature television tube you see in the cockpit.

The more modern and sophisticated the unit, the more easily it can be understood by the Right Seater. Radars can be set for certain ranges, and the pilot will often use a number of settings to get a picture of what is happening in the near, intermediate, and far ranges, below, ahead, or above. By looking at the unit, you'll be able to see what range the pilot has set in. If it's set, for example, for 25 miles, the screen will show a 25-mile arc and perhaps other arcs at intervals.

Primarily, the pilot uses radar to see precipitation, evaluating the picture to discern whether it's just an ordinary rain shower, or the heavy rain and severe turbulence associated with a thunderstorm. Navigation is a secondary use for radar, to read very clearly defined terrain such as shorelines, small islands, or mountains.

At first, the screen won't make a lot of sense to you in the Right Seat. The first key is to note what range the radar is set for. The rain you see with the naked eye might be too far away to see on radar at first. As you approach, the pilot may select closer ranges, making the area look bigger and scarier to the non-pilots aboard. By examining the area in more detail, the pilot can pick out its soft spots or select a way around it.

If you know your ground speed, note on the scope whether the radar is set for 25 miles. If all the weather is within that picture, you can easily calculate in your head how many minutes it will take before you're out of the rain.

If it's a color radar, it can differentiate among rates of rainfall more easily. More sophisticated units, used in airliners and military planes but unlikely to be found in private airplanes because of their high cost and heavy weight, can do far more that this but we hope you now feel more comfortable with, or without, radar aboard.

Descent and Landing

Well in advance of landing the pilot will start letting down. If there are any ear problems abroad, the descent can usually be planned to be lengthier and more gradual — depending, of course, on the terrain, temperature, turbulence and other flight factors.

Well before you're aware that the ground is coming up to meet you, you'll note changes in the instruments. The altimeter unwinds, the ROC shows the rate of descent in feet per minute, your ears may pop, and airspeed increases. In VFR flight, the pilot will look up an airport diagram or other description, will tune in the airport's Unicom or Approach Control, will reset the altimeter, and may change fuel tanks to use the fullest one. The prelanding checklist begins.

At this point, your help in watching for traffic is especially appreciated.

Be prepared for some alarming sounds. The pilot will probably check the Gear Warning horn and, just before touchdown, the stall warning horn will probably sound off, indicating that it's a proper, full-stall landing. Once reclaimed by the ravages of the runway, the plane rattles and groans almost as raucously as it did on takeoff.

As the landing roll ends, taxiing begins, often with the pilot doing the after-landing checklist at the same time. At a controlled airport, important instructions continue to come in by radio, so the pilot can't relax until the plane comes to a stop and the propeller is no longer turning.

Blessed silence descends, but there is one more task at hand. The pilot now logs the duration of the flight and how much fuel is remaining, and notes any items that should be repaired or looked into before the next flight. One last check is made to make sure everything, especially the magnetos, is turned off before anyone goes near the prop.

Once the airplane is secured — and this can be a long process involving tiedowns, sun shades, and control locks — and the flight plan has been closed, you have arrived.

About Night Flying

Everything looks different when you're flying at night, all black velvet and studded with jewels. Navigation is easier because communities can be seen from miles away. Busy highways show up as continuous lines of white and red. The flashing green and white beacons at airports can be seen from 50 miles away or more in clear weather. Coastal shorelines are precisely outlined. Moonlight reflects off lakes, rivers, and ponds.

The pilot may keep the panel lights so low you can barely see them, using a dim red or white light to read charts. It's important to keep interior lighting dim, to protect the pilot's night vision.

If there's a blackout curtain between the front and rear seats, it can be closed so passengers can use their reading lights. Never, never light a flashlight or a reading light in the Right Seat without permission because it could take a long time for the pilot's night vision to return to a level of sharpness necessary to pick out a line of feeble lights marking a small, dimly lit airport. Larger airports, of course, are a veritable carnival of light, so night vision here isn't quite as critical as it is in back-country flying.

Recognizing a Collision Course

Most of us have no problem recognizing a collision course in the two-dimensional world of the highway, but in flight things seem to happen in slow motion, making it difficult to gauge another aircraft's speed, altitude, and direction.

The Left Seat has many means of knowing what traffic is in the area and who is going where. Pilots hear other pilots on the radio; they know that standard flight procedures will result in certain movements of another plane: and their eyesight is usually fine-tuned to picking out other traffic. The chances of a near miss are very small, but help from the Right Seat can always make the chances even smaller.

If you see another aircraft directly on the horizon, that means it is at or near your altitude. And if it stays in the same relative spot while it continues to get larger, it could be on a collision course with you. When you see such a speck at any point ahead, behind, or to the side of the airplane, watch it for a few minutes to see if its position changes relative to a spot on your window or windshield.

Usually it appears to hang there and then it begins moving faster and faster away from your spot of reference. Even if it gets larger and passes near you at lightning speed, it will be clear to you well in advance that it will pass over, under, ahead, or behind your airplane.

At night, it's far easier to spot other airplanes because they're all sporting nav lights plus, in most cases, strobe lights and a rotating beacon. You'll see them sooner, but may have more trouble sorting out their size and direction. Again, the first thing to observe is whether the other plane is on your horizon line. Second, try to make out the nav lights, which will tell you what direction the other airplane is flying. Keep in mind that the green light is on the right wingtip, the red on the left, and the white is on the back of the tail, i.e. if you see the white light, you're behind the other craft.

The Cockpit Companion

The lighting placement is laid out by degrees, but it takes a pilot's practiced eye to interpret quickly which way the other airplane is going and how close it is. Unless the Right Seater has a lot of experience in boating or flying at night, about the only help you can give is to point to the target and let the pilot figure out the rest.

What's Going On Here?

Chapter 8

Health and Safety Aboard

In a single-pilot operation, the pilot's responsibility is to the flight. No matter how sick or scared a passenger may become, everyone loses if the pilot must divert his or her attention away from the single goal of getting the airplane and its occupants safely back to terra firma.

So, if the Right Seat can handle emergencies that are not flight related, everyone wins. In matters of health and safety, that means knowing the location of emergency equipment and being alert for any danger signs such as hypoxia or panic.

"Of course, half the battle is not to get into the airplane in the first place if everyone is not quite up to par," says Dr. Newell O. Pugh, director of the Cancer Center at Methodist Hospital in Indianapolis and himself a pilot with Commercial, SMEL, Instrument, and CFI ratings.

After an evening of partying, the designated driver might pour Goodole Charlie or Charlene into the back seat of the car and drive home. In an airplane, however, Charlie's intoxication will be increased with altitude, and your problems as Right Seat, housemother, father confessor, bouncer, and night nurse will be compounded by the fact that there is (1) no room for Charlie to dance with a lampshade on his head, (2) nowhere for the rest of you to go while Charlie tries to dance with the lampshade on his head, and (3) no way to evict Charlie at 6000 feet AGL.

As altitude increases, Charlie's behavior will intensify. In short, he'll get quarrelsome faster, pass out earlier, get horny sooner, or blubber more tearfully — and you're left to restrain him, console him, clean up after him, or otherwise deal with personality traits you never knew Charlie had.

Preventive medicine, i.e. staying at ground zero, also applies to the passenger who is too pregnant, or has emphysema or heart trouble or any other breathing problems, or who is in any way impaired or unpredictable, physically or mentally.

Pet on Board

A passenger can also present a safety problem if he shows up with his favorite dog or cat. (We also heard of a flight in which a pet turtle ran amok.) Although some pets are experienced flyers who have logged many peaceful hours aloft, Dr. Pugh would prefer that pets travel only when they can be controlled in a cage or at least with a muzzle. Again he says, "It's not like riding in a car." Like most other flying problems, potential pet problems are best dealt with on the ground, before takeoff.

Every pet owner has seen Fluffy or Rover revert to instinctive behavior at some time. Even the most loving dog will turn on you if it's hurt; the most peaceful cat will claw you if it feels threatened. And we've known both cats and dogs that experienced motion sickness. The last thing you need when a flight is uncomfortable or in trouble is a pet that wants out at any cost.

Fire Extinguishers

The first step is to know where fire extinguishers are stowed, and how they operate. That includes taking the unit out of its holder for a good look. The simple act of removing it can in itself can be more difficult, or require more strength, than you think.

If you fly regularly in this airplane, check the extinguishers before each flight to see that they read "in the green" and in date. If they are the Halon type, gauges can err. They must be weighed periodically to see if they still carry a full charge.

If they are the dry powder type, the gauge probably reads true. However, dry powder extinguishers carried in vehicles can fail to deliver a full charge because, when the cylinder is left in the same position month after month, the powder packs down from the constant vibration. Turn the extinguisher over occasionally, or give it a few whacks on the bottom.

When an extinguisher is due for recharge, it can be turned into a learning experience for the Right Seater and the entire crew. Call your local fire department and ask if there is a place where fire extinguishers can be discharged and recharged. (Don't shoot it off just anywhere; it is messy, corrosive, and dangerous.)

By discharging the extinguisher, you'll get a hands-on feel for how difficult it is to pull the handle, how far it shoots, how long the charge lasts, and what kind of spray pattern it makes.

Life Vests and Life Rafts

Aviation survival expert Wayne Williams now works as a consultant, but we once attended sea survival courses he was conducting at Nova University in Fort Lauderdale, FL. He found that, while most survival equipment worked as its manufacturers intended, it often failed because people failed.

In one experiment, he found that only a handful of people could don an airline-style life vest properly, even though it was daylight, on dry land, in a panic-free situation, and they had seen the cabin attendant demonstrate the vest dozens of times. Add darkness, panic, evacuation, crowding, and water to the equation, and you have a stickywicket indeed.

The same user-failure rate applied to life rafts, flares, lift slings, and other equipment that was itself in perfectly usable condition. He convinced us that there is no substitute for hands-on experience.

Don't be shy about looking into the packet from time to time, and reviewing the use of any life vests that are carried permanently in the airplane in which you ride Right Seat. Check for expiration dates, and have them repacked as necessary. If you're just renting the equipment for an over-water flight, ask for a sample and have everyone aboard try it on.

The wisest course, one followed by many people who fly the Bahamas' 700 islands, is to don life jackets before takeoff and wear them for the entire flight. In any situation where you're over water immediately after takeoff or before landing, there isn't time to find and get into life vests if there is an engine failure.

Don't use PFD's as beach toys. Some inflate automatically when wet, after which you're faced with a hefty bill for repacking. All PFD's should be kept clean, dry, and out of direct sunlight.

If you carry a life raft aboard all the time, it too will need periodic refurbishing. If it's out of date and has to be repacked anyway, this is an excellent time to give it a trial run.

Deploy it only in the water. It could be damaged if it's inflated on dry land because of the quick temperature change as the CO_2 cartridge goes off. Its materials and stresses were engineered for the water. Besides, the training session will be far better if you go through the entire procedure with both you, and the raft, in the water as you would be in a ditching emergency.

By inflating the raft, you'll learn how much pull it takes on the lanyard to start the process, how hard it is to climb into the raft from the water, and how stable it is. Practice a routine for pulling others aboard. If the raft flips, survival gear can be lost.

At this time, review the quantity and type of emergency gear packed with the raft and make sure it's updated at the same time the raft is. Read up on the uses of the sea anchor, solar still, EPIRB, and other survival gear stowed inside the raft.

The EPIRB (Emergency Position Indicator Radio Beacon), incidentally, is a marine version of the airplane's ELT, with important differences. While the ELT goes off automatically on impact, some EPIRBs are triggered manually and must be deployed in the water, which serves as the ground plane of the antenna. Once turned on, it must be left on. To turn it off in hopes of saving the battery defeats the purpose.

EPIRBs can also be tested, but only for one second during

the first five minutes of any hour, to avoid sending a false alarm. EPIRB batteries, like ELT batteries, are dated and must be replaced periodically.

If flares are carried aboard the airplane, these too go out of date and must be replaced. Again, this is a good time for a dress rehearsal in which the flares are actually set off. And again, this must be done under certain circumstances to avoid sending a false alarm. Contact local marine authorities, or groups including the Power Squadron or the U.S. Coast Guard Auxiliary, for information on times and places where flares can be fired.

One more point about rescue and survival situations. Once the professionals arrive at the rescue scene, they are in charge and it is important that their directions be followed. If you fly over water extensively, and a water survival course is available in your area, taking it is well worth the time and money.

Most commonly taken by professional pilots, seamen, and people who work on offshore oil rigs, such courses apply to any families who fly over water, and the skills taught also come in handy when you're boating, cruising, or even driving in areas where there is a danger of going into the water.

Kids and Motion Sickness

We may hold the world's Junior Olympic barf record. For many of the little Back Seaters we have had aboard, airsickness struck immediately and without warning. Poor little Age 9 retched miserably for an entire two-hour flight home, despite having taken a motion sickness pill well before we took off.

Children (a) vomit easily anyway, and (b) are particularly susceptible to motion sickness in small planes. So, be secretly prepared to deploy an airsick bag, then be pleasantly surprised if it isn't needed.

Having suffered our share of mal de mer while sailing, we ache for anyone who suffers from motion sickness. We learned at sea that two of the best defenses are plenty of fresh air and a clear view of the horizon. It's our guess that many children get airsick simply because their view of the flight, from deep in their seats, is one of swaying curtains and careening clouds. If they can be seated high enough to see the horizon, so much the better.

Arthur Hankin, the Philadelphia attorney and a Cessna Skylane owner who has been flying his children since they were infants, tells us his daughter was airsick only once. "It was a hot, bumpy day, and she had her head down in a book," he remembers. It was a lethal combination, one to be avoided.

"I recommend flying on a full stomach," he says. "Just don't eat too much or foods that are irritating."

Talk to your pediatrician about medications, especially for motion sickness and for ear problems, which plague children more than adults. Ask too about the best way to administer the medication. Motion sickness pills are tough enough for some adults to manage; for children, a liquid or suppository may be preferable.

The key to any such medication is that it must be administered before the airplane leaves the ground — usually at least 30 minutes in advance. (Discuss it with both the physician and the pharmacist.) Once the child starts feeling sick, it's too late for the drug to work, even if it stays down. However, it will probably be the first thing to come up — which is why, in protracted vomiting, such medications are administered in suppository form.

That Ol' Devil Prop

Imagine an almost invisible saber shearing the air with incredible power, and you can imagine the danger presented by the airplane's propeller(s). As we have said before, everyone connected with the flight needs to know two things. First, propellers are a clear and present danger any time the engine is running. Second, propellers are a clear and present danger when the engine is not running.

Bottom line: stay out of the arc of rotation at all times, and stay well clear of any aircraft whose engine is running.

We may think we're alerted by the noise, but at an airport one becomes almost inured to the intense and constant din to the point where it's hard to keep track of which airplanes are

The Cockpit Companion

Sometimes it's necessary to work within the propeller arc, but it's crucial to know when and how.
(Photo courtesy of Embry-Riddle Aeronautical University)

running. Not only children and pets, but longtime airport regulars continue to be injured in propeller accidents. Stay clear.

First Aid

The best first aid kit is the one you assemble yourself for your own airplane according to the most likely needs of your own flying companions. The airplane probably has a nice, neat box somewhere aboard with a red cross on it, but don't assume it's complete, or even usable. It needs modification now for your own needs, and frequent refreshing throughout the year.

Supplies get used up and dried out, or deteriorate in the temperature extremes of a stored airplane. Too, you may want to fine-tune the kit for each occasion, adding a snake bite kit for a hunting trip or some extra liniment for the ski weekend. The kit you have aboard for short local hops will be different from one carried for longer, overnight trips.

One of our favorite commercial first aid kits in the compact category is the small, very basic Family Pack made by Sawyer,

800/940-4464. It comes in a zippered, bright red, soft pouch the size of a paperback novel. There is some extra room inside to add things that are important to your mode of travel.

Its contents make a good checklist for any minimal kit: 5 yards of self-stick pressure wrap, gauze and bandages of various sizes and types, moleskin, a reusable forehead thermometer, tweezers, needle, safety pin, electrolyte tablets, triple antibiotic ointment, non-aspirin pain relief, antiseptics, a first aid manual with each state's emergency telephone numbers, and pill vials for carrying individual prescriptions.

For truly flyweight travel, Sawyer makes a small pack that weighs only a few ounces and has belt loops so you can wear it anywhere. It contains 5 yards of wrap, bandages and gauze pads, antiseptics, antibiotic ointment, electrolyte tablets for heat exhaustion, non-aspirin pain pills, and a first aid manual with state by state emergency phone numbers.

Among items you may want to put together for family travel are all the above items in whatever type and quantity you can manage, plus:

- Proof of medical insurance, policy numbers, and telephone numbers for your physician, pediatrician, pharmacist, poison control center, and (if you travel with pets) your veterinarian. This sounds like a long list, but it will fit on the back of a business card. Also, take written copies of prescriptions, preferably by their generic name. Prescriptions aren't always honored by pharmacists in other states, but with the written order, and telephone numbers where the prescription can be verified, you are almost certain to get the medication you need.

- Sun block, lip balm with sun screen, insect repellents, and tick repellent. Until Lyme disease is licked, it's important to be protected against this tick-born disease. Most ordinary mosquito sprays do not repel ticks, which are now found in almost every state.

- We consider sunglasses so important in the cockpit, we always have at least one extra pair aboard among our emergency supplies.

- A small, inexpensive magnifying glass is worth its weight in gold if you have to remove a sliver or tick, or read microscopic instructions on a medication.

The Cockpit Companion

- Anaphylactic shock kit, especially if anyone aboard is sensitive to stings. Small airports, especially grass fields, are alive with insects.
- Topical, "anti-itch" treatment for minor insect bites.
- At least one each disposable heat and ice packs that can be activated as needed.
- Elastic bandage(s).
- Over-the-counter medications for travelers' diarrhea, constipation, motion sickness, pain, heartburn, sniffles, and coughs. By buying sample sizes, you can have a comprehensive pharmacy of fresh, packaged, well-sealed medications for only a few dollars.
- Antifungal powder comes in handy if you're prone to pick up athlete's foot.
- A small tube of topical anesthetic for sores in the mouth and teething.
- Petroleum jelly. Among its other uses, it can be used to reinstall a crown that comes off a tooth. Seek help from a dentist as soon as possible; this is a temporary fix. Permanent adhesives should not be used for two reasons. One, you could re-cement

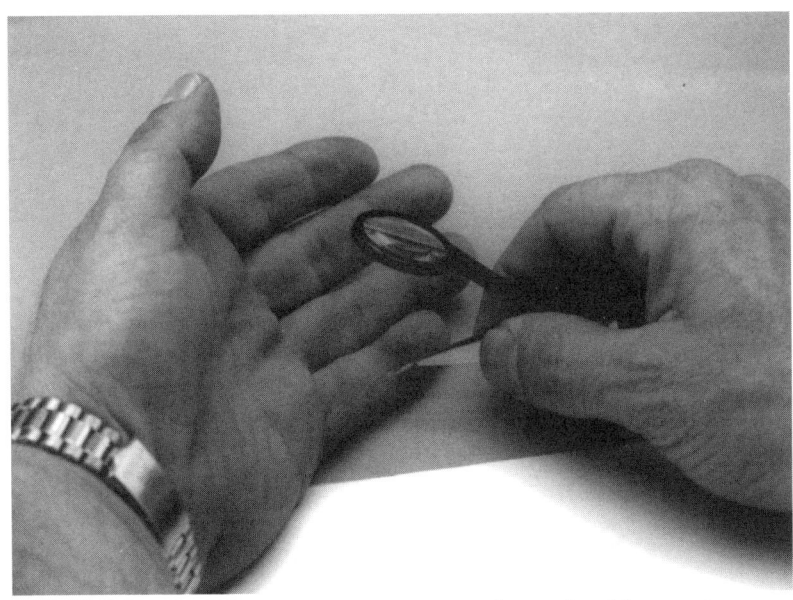

A magnifying glass takes up little room in the medical kit, yet it comes in handy for many tasks. This model incorporates tweezers, which are also a first aid basic. (Photo courtesy of Miracle Point)

the tooth off center, creating a new "bite" which in turn causes problems in the jaw hinge. Two, the crown may have loosened because of decay underneath, so it's folly to cement the crown on atop the decay. A permanent, professional repair will be needed.

• A few tea bags. A folded tea bag, soaked in warm water, can be placed over the socket of a tooth that was knocked out and lost. (If the tooth is available, it can be replaced. Ask your dentist for a brochure giving step-by-step instructions for this. The sooner you act, especially with a child, the better the chances of re-rooting the tooth.) Wet tea bags are also a folk remedy for sunburn.

• Artificial tears to refresh tired eyes.

• Sturdy zip-top plastic bag, such as the Sure Seal brand. If you keep at least one in the first aid kit, you'll always have a clean bag on hand for making an ice pack.

Also handy to have on hand:

• Individually wrapped bar soaps and a small jug of water. Any first aid treatment begins with clean hands.

• Ice. It is first aid for so many injuries and ills, including nose bleeds, it is always advisable to have a small ice chest aboard.

• Ipecac syrup for inducing vomiting in children who have swallowed toxins. Make sure you understand, however, which toxins should not be treated this way.

• A flashlight or a chemical light stick. Although a flashlight is aboard for general emergency use, it's handy to have one bought for and kept in in the first aid kit.

About SCUBA Diving

As part of your certification as a SCUBA diver, you've already learned the guidelines that apply to flying after a dive. (If you're not certified, don't dive.) Remember that these guidelines apply to travel in small airplanes too. Decompression sickness is an emergency. The most common symptom is pain in the joints and abdomen. It is because the pain causes victims to double over that it's called the "bends." Itching or a rash may also occur. Radio ahead for help, and get to terra firma ASAP.

If You're Asked to Prop a Plane

Because antique and restored airplanes are becoming more popular all the time, and many of them have not had electrical systems and starters added, "propping" is still very much a part of the aviation scene and the Right Seater is the logical person to be pressed into service.

Don't do it. Not until you have had a complete and thorough introduction, supervised by a skilled propper, to the art of hand starting an aircraft engine. It's not a matter of brute strength, but is more an exercise in balance, ballet, acrobatics, and Looking Out for Number One.

Instruction begins with an engine that is stone, cold dead. Before you — as a novice propper — approach the prop for your first lesson, the pilot will see that all the high tension leads to the spark plugs are removed. Remember there are two spark plugs in each cylinder, and make sure all are off.

In a car, all the wiring has to be perfect before you get spark. With a magneto, however, a ground wire that is *not* perfect can allow the magneto to fire a spark plug. A magneto is turned off only by grounding it out through the ignition switch, which is On when it's Off. An airplane almost always has two mags and, unless they are tested for Off during runup, it's very possible that at least one will be hot.

All it takes is one misfire to give you a manicure up to the shoulders. The gas selector valve should be in the Off position, and the throttle should be completely closed. Only when the engine has been completely disabled should instruction begin.

The object is to pull the engine crisply through at least one compression stroke while, at the same time, maintaining a stance in which the flow-through motion takes you away from the propeller no matter what else happens. Practice at least 25 starts; 100 wouldn't be unreasonable, even if you have to go home to the heating pad and come back the next day to continue the course.

Repeated cranking does not hurt the engine. Before the real thing, you must be absolutely certain that you're in control of your own movements.

When everything is hooked back up and you are actually starting the engine, the routine will be different for each airplane, but by then you'll be well prepared — both mentally and athletically, for both the expected and the unexpected.

One more caveat. Having propped a J-3 does not prepare you for propping a Belchfire Zephyr. Leave any airplane with a very large engine, or a non-standard ignition system to the experts. Don't, for example, try to hand prop a Cessna 195, which has one distributor and one mag, and can fire when you least expect it.

The usual rule of thumb is that any number of cylinders above four becomes very difficult, and any number of blades more than two becomes deadly.

About Hyperventilation

If a passenger becomes panicky and begins breathing quickly and shallowly, gulping in great quantities of air, dizziness and disorientation result. Hyperventilation causes an imbalance in the ratio of oxygen to carbon dioxide in the blood. You may have experienced it yourself after blowing up too many birthday balloons.

If you can find a paper (not plastic) bag, suggest calmly that the person hold it over their mouth and nose and breathe into it. Don't force the issue or you may aggravate the panic. Balance is restored to the blood as the victim breathes back the carbon dioxide that had been exhaled.

Hyperventilation is not to be confused with hypoxia, or lack of oxygen. This can results in pale skin and dizziness, and may require a few puffs from a portable oxygen bottle. It shouldn't be a problem with healthy people at altitudes below 10,000 feet.

The Eyes Have It

The single most useful thing that the Right Seater can do to aid in the flight is to provide another pair of eyes. Are yours all they can be?

The Cockpit Companion

Nowhere do eyes work harder than in an airplane cockpit, where focus is constantly changing from infinity to the instrument panel, to the chart, back to the panel, then off to the side to look for reported traffic. Experienced pilots develop a certain radar for spotting traffic, not just in general but in learning to focus on specific distances.

Reported traffic one mile away at two o'clock? Looking for the runway that you know to be three miles ahead? Inevitably, the Left Seat will spot it first because of what we believe is a sixth sense, developed through experience, that focuses the eyes for the approximate distance desired. Pilots and other observers who specialize in search and rescue tell us they also develop an inexplicable ability for spotting tiny anomalies in the endless, trackless, featureless ocean or over unbroken miles of look-alike forest.

In no other facet of life are eyes asked to do as much as they do in the cockpit. There, eyes labor under light conditions ranging from barely there to blindingly bright, at a variety of distances under ever-changing circumstances. If you believe in exercise, it comes as no surprise then that ophthalmologists have found that pilots' vision remains better longer than the population as a whole.

According to Richard Leviton, who wrote Seven Steps to Better Vision (Talman Company, 150 Fifth Ave., New York NY 10011) the steps to natural improvement of vision are:

1. Exercise the eyes to enhance their ability to relax, focus, shift, work as a team, and visualize.

2. Take advantage of the tools and techniques of behavioral optometry to optimize your ability to see with mind and body.

3. Become aware of how hidden emotions and basic psychological attitudes may hinder vision, and practice some of the mental exercises that can help you overcome psycho-emotional obstacles to better vision.

4. Learn to free the body's energy channels and treat minor vision problems using natural remedies.

5. Become conscious of how movement, posture, and alignment of neck, head, and torso can affect vision, and use exercises and adjustments to correct patterns that can impair better vision.

6. Feed your body and eyes a diet of foods, supplements, and light for optimum visual health.

7. Learn and adopt the positive lifestyle and work habits necessary for the long-term health of the visual system.

It takes Leviton an entire chapter to explain each of these steps fully. The book is an eye opener, well worth reading and discussing with your own eye care professional.

CPR

Any Right Seater who knows CPR can be the life of the party — literally. The course we took required only a few evenings; updating a certificate takes even less time once a year. Although CPR instructions are available in written form, there is no substitute for practice on a dummy, especially in infant CPR where a very light touch must be used. If you've taken this course in the distant past, take a refresher. Not only will it bring your certificate up to date, it will teach you all the new things that have been learned about CPR since it has come into widespread use.

Odds are that an emergency would not occur in flight, but if a passenger does need CPR it's hardly something a pilot could do while flying the airplane at the same time — even with an autopilot. Special techniques can be used in both CPR and the Heimlich maneuver when sitting down, in close quarters. Learn them too.

Once certified, the Right Seater who knows CPR can help save the day any time during any fly-in vacation or business trip, by giving the breath of life at the scene of an accident, drowning, or collapse.

Another first aid technique that is easily learned is how to replace a dislocated jaw, a problem that can occur from something as simple as yawning. Ask your dentist.

Chapter 9

The Knacks of Snacks

Snacking, Packing, and Stewarding

In our family, Grandpa always had a pocketful of secret surprises for any of us who got bored, restless, hungry, queasy, or had a coughing spell. His all-purpose remedy was white peppermints, and they worked wonders. In the airplane, it is only natural that such solaces are in the charge of the Right Seat.

Olympic gold medal winner Bruce Jenner and his wife Kris travel often with their children in their MU-2. Although their flights are usually kept short and sweet for the children's sakes, the Jenners sometimes take sandwiches aboard, and always have plenty of individual juice boxes. "Just be careful when you open them," Jenner warns. Altitude can do funny things to liquids, even if they are not carbonated.

Dining aloft isn't always easy, but it is a good idea to have some food and drink aboard if only for emergencies. Bottled water, sealed jars of unsalted peanuts, and canned prunes are among those items that can be left on board for months.

Conversely, it isn't wise to leave anything aboard that is wrapped only in plastic or paper. After we found that a half package of Lifesavers had been raided by an ingenious mouse who climbed the nose gear and made his way through a labyrinth of pathways and cables, we have been careful to remove all food from the airplane and to vacuum any crumbs. There's nothing like a scampering roach or, worse still, a curious mouse, to create a diversion during a flight. Pest control is a must, and it begins with removing every morsel of food after each trip.

At its best an airborne picnic eaten at 6000 feet, while looking down on golden wheatfields or hazy blue foothills, can be a feast for the eyes as well as for the palate. No royal banquet can compare with a picnic lunch eaten next to a grassy runway under the shade of a wing. Or a basket of breakfast croissants and a thermos of steamy coffee while watching the sunrise before an early takeoff.

Here are some ideas for the Right Seater who doubles as Senior Cabin Attendant:

• For a taste treat: Rolls of Lifesavers, individually wrapped hard candies, Tic Tacs, and chewing gum (but don't give it to children who aren't old enough to dispose of gum responsibly.) Jawbreakers, all-day suckers, and other large hard candies last and last, stretching out the pleasure and diversion.

• Whether it is breakfast, sandwiches, or snacks, make up individual meals in sturdy, plastic, zip-top bags. Even if you'll be buying foods along the way, rather than bringing them from home, have a good supply of empty plastic bags on hand. We prefer the freezer type bags because they are sturdier. Leftovers

and trash can then be sealed in these leakproof bags.
- Individual servings, as the airlines know so well, are the key to quick, neat serving in tight quarters. Individually wrapped "lunchbox packs" of potato chips, corn chips, cheese or peanut butter snacks, and cookies cost more per serving, but are neater, more sanitary, and they stay fresh until opened. Lunch pack juices and milk stay fresh until opened. They come with their own straw, and are almost completely leakproof. No matter how many sandwiches it takes to make a meal, wrap each individually. Better still, wrap each half separately.
- Don't underestimate the filling, nutritious, comfort food value of goodol' bread and butter. Start with the freshest and best breads and good butter. Butter both pieces evenly with softened butter, then make into sandwiches and cut into halves or quarters. Seal in zip-top sandwich bags.
- Apples, seedless grapes cut into "bunchlings", or firm pears are the best fruit to carry. Bananas smell up the cabin; berries are too fragile; almost all other fruits need to be peeled, cut up, seeded, or otherwise wrestled into submission.
- Sticky sandwiches are less likely to drift apart. Good choices for eating-on-the-go are egg salad, chicken salad, ham salad, and peanut butter and jelly. In hot weather, the safest choices without refrigeration are peanut butter and jelly or cheese with mustard. If you'll be eating in your seat, forget messy additions such as shredded lettuce, sliced tomatoes, or sprouts.
- When making sandwiches, avoid breads that are sliced extra thin, pita (our pitas always leak at the seams, especially if the sandwiches were made ahead of time), and breads that have tough, chewy or brittle crusts.
- To keep canned drinks cold without carrying a cooler, place each can in an inexpensive foam can insulator, then clamp a second insulator on top.
- To make packaged fruits and vegetables more crushproof, seal them in zip-top bags with a good cushion of air.
- Although plastic sandwich bags are popular, the advantage to wrapping sandwiches in waxed paper is that the paper opens out to serve as a lap cover and catch-all.
- Don't forget recycling, even when aloft. All it takes is a few extra bags to separate aluminum and glass from paper and plastic.

- Pack an individual hand wipe with each meal, and have plenty of extras on hand.
- Don't fuss at children about eating. Lorraine Miller, professor of nutrition and food services at Oregon State University says, "When it comes to hunger...kids seem to have their own body clock, and they are very good at listening to it." Researchers working with children ages 2-5 found that, left to their own devices, children maintained a relatively constant daily caloric intake, even though calorie counts varied greatly from meal to meal. Keep graham crackers, unbuttered popcorn, fruits, vegetables, and yogurt handy.

Here are some menu ideas

Nosebag lunch #1 (Keep cold)
Large zip-top freezer bag containing:
1 cranberry juice cocktail with straw
One or more chicken salad sandwiches. Combine chopped chicken, salad dressing, finely diced celery, broken cashews, and well-drained crushed pineapple. Use buttered whole grain bread.
1 packet potato chips or individual bag carrot sticks
1 Golden Delicious apple
Individual dessert (granola bar or snack cake)
1 stick peppermint chewing gum for a refreshing wind up to the meal
1 packaged hand wipe
1 large napkin

Nosebag lunch #2 (No refrigeration needed)
Seal in each zip-top bag:
1 carton apple juice with straw
1-3 individually wrapped cheese wedges
1-2 packets individually wrapped saltines
1 cluster seedless red or white grapes
Lunchbox pack individually wrapped brownie
Roll peppermint or wintergreen Lifesavers

The Cockpit Companion

1 hand wipe
1 large paper napkin

Breakfast Business Briefing

Take this one to the airport. Get some of the pre-flight preliminaries out of the way while you work up an appetite. Then buy or mooch cups of airport coffee and pass out these individual breakfasts before boarding. If you can find tiny, individual salt packets, add one to each. If not, carry along a small, disposable container of salt to be shared by those who like to salt their hard-boiled eggs.

In each bag, seal:
1 can peach or pear nectar
Individually wrapped straw
1 or 2 big, fluffy baking powder biscuits, split, buttered, and pasted with paper-thin slices of cold fried country ham.
1 or 2 hard-boiled eggs per person, peeled, and wrapped individually in plastic wrap
1 packet mint candies (for now and later)
1 large paper napkin
1 hand wipe

Romance at Sunrise

Take a cue from hot air balloon operators, who take paying passengers floating into the morning sky, and put down in a meadow for a champagne picnic. Pick a small, scenic field half an hour or so from your home airport. Take off at sunup, enjoy the flight, then land and share this breakfast on a velvety beach blanket spread under a wing.

In a small ice chest, chill a bottle or two of sparkling, non-alcoholic apple or grape juice. Pack big, colorful paper napkins, stemmed plastic wine glasses, and a couple of long-stem pink roses. Add to the menu:

Croissant sandwiches. Split big croissants, spread both cut sides with softened cream cheese, then add a layer of sliced strawberries, or layers of lox and thinly sliced onions, or strips of crisply fried bacon. Wrap individually.

Big bran muffins, sweet with honey and plump with raisins. Split, butter, and wrap them individually.

Cartons of creamy yogurt with fruit. Don't forget disposable spoons.

Extra fancy roasted, salted nuts. Put them on a square of plastic wrap, and tie with a ribbon.

Raspberry tea. Bring it in a thermos and serve it as a last course to the meal.

Add a small packet of heart-shaped cinnamon redhots to refresh the mouth after eating.

For a tingling clean-up, wring out a couple of pastel wash cloths in plain water, then splash them lightly with an herbal scent. Zip them individually in plastic sandwich bags.

Dawn Patrol

This is a brawny breakfast for the hunting or fishing party. If it's to be eaten aloft, it's always best if you can manage individual thermoses of coffee. To the list below, add disposable cups if you'll be serving from a large thermos.

Pack in each plastic bag:

Individual can extra zippy vegetable juice cocktail

Individually wrapped straw

1 or 2 fried egg and crisp bacon sandwiches on beef steak rye bread that has been buttered then spread with dark mustard. Fry the eggs individually, scrambling to mix and shaping as best you can to the size of the bread. Cook thoroughly. Make the sandwiches the night before, and refrigerate them until morning.

1 individually wrapped streusel coffee cake (commercial versions are available in "lunchbox" packs)

1 stick cinnamon gum

Large paper napkin

1 hand wipe

Now wrap each bag in a big bandanna kerchief and knot it hobo style. After breakfast, trash can be corralled in the plastic bag. The bandannas, tucked into guests' pockets, go along on the hunt, where they'll come in handy as kerchiefs, sweat bands, or hand towels. Send them home with your guests, as souvenirs of a great day's fishing or hunting.

The Ultimate Family Fly-In Picnic

To the aviation family, the fly-in picnic is a modern version

The Cockpit Companion

of the old Sunday afternoon ride in the country. Pick a scenic cross-country route, put down at a country airport, spread a picnic, then take a snooze or play a game of frisbee before flying home again.

Pack a ground cloth, a folding umbrella or a tarp that spreads over the wing to provide shade, and a can of yard fogger for bugs. We also carry a couple of flyweight, compact, legless folding chairs. Our ice chests are the soft type, which can be squeezed into odd shapes to fit anywhere we have room.

Don't forget hand wipes, napkins, straws or drinking cups for the cold drinks, and bags for trash. With this menu, you can eat from the lap without paper plates or any utensils except for those listed.

The Menu:

Individual cans or boxes juice, milk, or soda.

Individual hero sandwiches. Split and butter whole grain hero rolls, and spread thickly on cut sides with butter, then salad dressing or a grainy mustard. Allowing for individual tastes, make up sandwiches using thinly sliced provolone and swiss cheese, salami, breast of turkey, sweet onions, rings of bell pepper, and sliced pickles. Wrap each individually, keying them with colored plastic wrap if necessary so you'll know that the pink one is Susie's-hold-the-onions, blue is for Hank and Biff with double on the salami, and green for just plain heroes. Make them the night before, and refrigerate. Then pack them in an ice chest with one or two "blue ice" containers.

Hot Bouillon. Slice one or two lemons paper thin, and seal the slices in a small plastic bag. To serve, place a teaspoon of instant bouillon granules in each cup, fill with boiling water from a thermos, and float a lemon slice on top. It's sipped, not spooned, so you don't need utensils for everyone. Although bouillon cubes can also be used, they take longer to dissolve and require stirring. Granules blend instantly with the hot water.

Crisp red apples. Wash and dry the apples at home, and wrap each separately in a clean paper napkin to cushion it and keep it clean. The extra napkins will come in handy.

Chocolate cupcakes. Buy individually wrapped, cream-filled commercial goodies. Or, if you'd prefer a homemade dessert, bake chocolate cupcakes in individual, flat-bottom ice cream cones. Just before serving, break out a can of vanilla frosting and a spatula, and swirl them with frosting as you pass them out.

About Packing

Packing for travel in small planes can be a challenge, even to the most experienced backpacker. The smaller the luggage area, the bigger the challenge. Yet some of the best flying trips are those in the smallest airplanes. Here are some observations based on our travel experiences.

• Lightweight, fabric duffels are a must but keep in mind that you'll eventually have to load or unload luggage in the rain. You may want to treat the duffels with a couple of coats of water repellent spray and a seam sealer.

• Keep a half dozen or so of the largest size plastic trash bags on board. Use them to wrap luggage that has to sit out in the rain. In a pinch, you can turn them into ponchos for yourselves too.

• Learn to pack Navy style, by folding clothes carefully and then rolling them tightly. They'll travel more wrinkle free, even though they're packed more tightly.

• Think small. Full featured, but very small electric razors, travel irons, and hair dryers are available through specialty catalogues. So are folding motorbikes, inflatable boats, telescoping fishing rods, and space savers for almost any other travel interests.

• Start a collection of underwear and socks that are good for just one more wearing before they're thrown away. Take them on your trip and, rather than carrying a growing collection of laundry, discard them as they are worn. If you travel with a baby and use cloth diapers, use the same system. Take stacks of clean, but nearly worn out, diapers and discard them (responsibly, of course) as you go.

• Take a small, heavy duty plastic zip-top bag filled with concentrated laundry powder, and a couple of plastic hangers. (Most hotel hangers can't be removed from the closet.) If you do your own drip-dries, you can get by for weeks with only one or two outfits.

- Instead of rolling clean socks, use each sock as a stuff bag for a rolled up sweater, hat, down vest, or other item that needs compressing.
- An unlined nylon windbreaker, the kind that folds up into itself and ends up the size of a wallet, is worth its weight in platinum. Ours came from L.L. Bean. Over a light sweater, a windbreaker is as good as a coat; over a sweat shirt it's warm enough for light winter weather.
- If you're packing for optimum warmth in the smallest space and weight, get hooded jackets, sweaters, or sweat shirts. No heat is lost around the neckline.
- No matter how informal the trip, each person should have at least one extra pair of shoes. Shoes are bulky, but can be tucked into odd spaces around the cabin if space is severely limited. Carry a small empty duffel to collect up all the shoes before leaving for the hotel.
- When you're really pinched for space, wear as many of the clothes as possible. Even in hot weather, you'll be cool enough once you're aloft. With a good bush jacket or fishing vest you can wear almost everything but the kitchen sink.
- For vacation travel, get acquainted with the lava-lava, the all-purpose garment worn in the South Pacific by both men and women. A plain length of cloth, it can be used as a bathing suit coverup, shawl, skirt, burnoose, bed cover, sarong or, with some judicious tucking and draping, pantaloons. Many tropics travelers get by with little more than a couple of these 3-yard pieces of light, 45 or 36-inch-wide cotton yardage.
- Consider carrying one or two disposable coveralls. Made from Tyvek, a papery plastic, they can keep you warm and dry if you're caught in unexpected cold, and clean if you have to do a dirty repair job in your good clothes. Cost is only about $5.
- Don't waste cabin space on pillows, booster cushions, or back supports. Get empty pillow shams, and fill them with things you want to carry along such as lap robes, extra jackets, tarps, or down vests. Or, get inflatable pillows that disappear when they're not in use.

The Knacks for Snacks

Chapter 10

Starring Roles for the Right Seat

So far, we've been talking about roles in which the Right Seat plays second fiddle to the pilot. However, many Right Seaters have special skills that the pilot does not have, skills that can complement pilot's abilities and make for a much stronger team.

Here are some areas in which the Right Seat's skills can be useful to the flight.

Camping

In all of its forms from backpacking to recreational vehicles, camping is an equal opportunity family sport. Add an airplane to your camping equipment list and you have the best of both worlds. Fly to a remote island in the Bahamas and camp on the beach. Land a float plane off Fort Jefferson in the Dry Tortugas off Key West and pitch a tent in the shadow of an enormous, Civil War-era fortress. Fly to EAA's Sun 'n Fun every winter in Lakeland, Florida, and party with the fly-in camping crowd.

Although some fly-in campers sleep in the open or in the airplane itself, the best camping set-ups combine a completely enclosed, screened, floored tent with a tarp that goes over a wing to provide a second "room". Just add sleeping bags and an

Camping can be fun for the whole family.

inflatable mattress and you're home.

Fly-in camping requires plenty of advance planning because the supplies you take with you will vary according to item availability at your destination. In the most primitive camping, you'll have to take all your needs including water and a folding shovel for digging a pit toilet.

Some airports have toilets and hot showers for fly-in campers. If there is a coffee shop, meals are available too. Many fields have rental cars available, and some small FBO's provide loaner cars so you can get to a restaurant, food store, or the local tourist attraction. Call ahead to get the current scoop on facilities, local transportation, and restaurant hours. Guidebook listings may be woefully out of date.

Computer Savvy

Pilots get most of their pre-flight information in one of three ways. They can visit a Flight Service Station (FSS) if the airport has one, telephone the FAA's toll-free weather briefing system, or call one of the private services that are available on a subscription basis.

Now there's another way. The pilot can plug into the DUATS

(Direct User Access Terminal System) to get a complete weather picture on a home computer. If you're a computer guru and your pilot is not, here's where you can really shine.

Only about half of today's 700,000 pilots have access to a computer; of those who do, some are still intimidated by them according to DUATS purveyors. Yet, to anyone who has computer skills — and this includes most kids over the age of ten — DUAT is a lead pipe cinch.

The menu starts with a choice of route and local weather briefings. If the pilot wants to continue with the menu and file a flight plan, this too can be done via DUATS. Even if you use the service only as a preliminary before leaving for the airport or before filing a flight plan by telephone, it can provide important flight planning help.

Basic service is available to anyone who has a computer capable of emulating a dumb terminal, plus a modem and communications software. For more sophisticated graphics, you'll need an IBM or compatible computer with at least 512K RAM, Dos 2.0 or higher, and an EGA or VGA monitor with graphics capability. Contel's Golden Eagle software is also available for MacIntosh. If you want to be able to print out the information, you will, of course, need additional equipment.

The power and convenience of the system are impressive. With Contel's DUATS, for example, you can file a flight plan just as if you called FSS directly — amend it, or cancel it up to one hour before departure (30 minutes for VFR flight plans). Weather is available in plain language, if you like, after it's delivered in FAA weather pilotspeak.

With your help with the computer part of it, the pilot can designate a route by Victor, Jet, or Area Navigation routes, NAVAIDS, SIDs, STARs, location identifiers, latitude/longitude, airports, fixed radial distances, or any combination of these, to any destination in the United States including flights into Canada or Mexico, and DUATS provides alphanumeric weather information along a 50-mile corridor through the entire route.

Start by contacting the following suppliers to see how you can log on. Then, if DUATS proves to be a useful tool in your Right Seat role, investigate optional extras.

GTE
Contel Federal Systems
Box 10814
Chantilly, VA 22021
Computer line (800) 767-9989
Help line (800) 345-3828
   Contel's Customer Service Center, (800) 345-DUAT, also answers 24 hours a day.
Or
Data Transformation Corp.
559 Greentree Rd.,
Turnersville, NJ 08012
Computer line (800) 245-3828
Help line (800) 243-3828

   Both of these services also offer value-added software at very modest prices. Additional software is also available from Flight Data Centers, 34 S. River St., Wilkes-Barre, PA 18702, (800) 451-3282. Ask about DUATS/Plus.
   DUATS service is free except when you're using value-added services, such as the Accu-Weather Radar Plus map like those seen on television weather shows, or flight planning services. Connect charges for the extras are modest, at about 30 cents per minute.

Other Computer Fun

   Computer bulletin boards abound, and you might plug into a few to see if one is of special interest to you and your pilot. If not, you may even want to start one for your particular interest such as fly-in fishing or hunting, fly-in tent camping, flying Alaska or the Bahamas, seaplanes, homebuilts, radio control model airplanes, flying doctors, flying farmers, flying lawyers, flying with infants or toddlers aboard, and so on.
   The FAA maintains a BBS (Bulletin Board System) that serves as an open forum among all users of the system including pilots of all experience levels, as well as FAA professionals. On it you'll find announcements of air shows, seminars and pilot

meetings; the names and numbers of FAA contacts; names of examiners in your area for medical, written, or pilot tests, and lots of other useful information to be exchanged among pilots, and between pilots and the FAA. Part of its role is to provide feedback from the flying public to the agency, 7 days a week, 24 hours a day.

Although some of the interchange is of national interest, much is local in scope so it's best to connect with a bulletin board in your own FSDO (Flight Standards District Office.) For information on the FAA-BBS nearest you, write the Federal Aviation Administration, Aviation Standards National Field Office, Box 25082, Oklahoma City, OK 73125.

Another FAA on-line computer information clearinghouse is FEDIX, which carries aviation news of all kinds, as well as information on education resources for teachers, and for students of all ages. Call up the menu by dialing (800) 232-4879. The Help line is (301) 975-0103. See Chapter 11 for information about NASA's education resource database.

Ham Radio

Like computers, ham radio opens access to a whole new world of communication that can enrich an aviation outing. If you're a ham, and if much of your flying is into remote hunting or fishing lodges or other areas that do not have good telephone service, your home ham radio shack is the key to all sorts of pre-trip arrangements, shortcuts, questions and answers, as well as to post-trip follow-ups.

Mechanics

Virtually all work on airplanes must be done by A&P licensed mechanics. Any electronics work has to be signed off by FCC-licensed technicians. Still, there are ways the mechanically inclined Right Seater with a tool kit can save the day for an all-thumbs pilot.

Keeping in mind that you must travel light, here are some

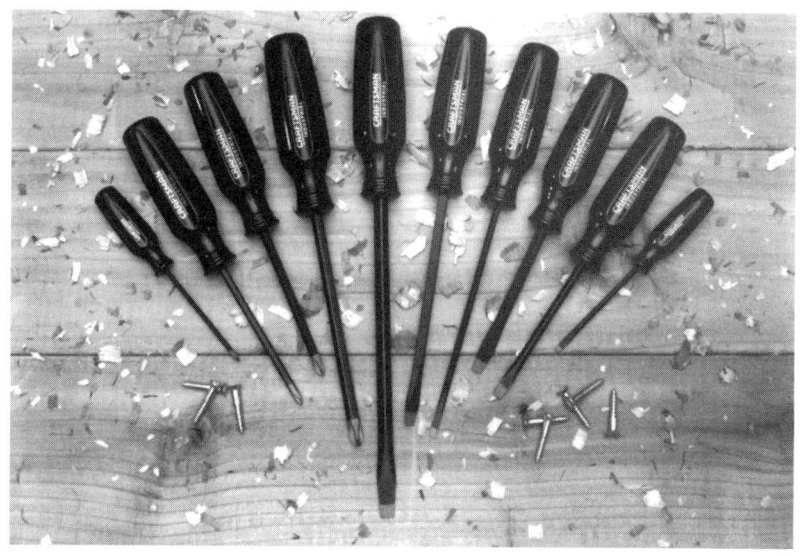

Keep your tool kit away from the compass. Some tools become magnetized and will offset your compass.

items that a basic tool kit could include. Don't keep the kit near the compass; tools sometimes become magnetized.

Multi-meter

We prefer the old analog type, one that can perform some of its functions without a battery. If there is any electrical failure aboard, you can check the battery and master switch relays and perhaps narrow down the source of the problem before calling in a technician. Many components, such as panel-mounted radios, can easily be removed from the airplane and taken to the service technician. Usually, it's easily done with a screwdriver. Even if the unit has to be reinstalled by a licensed professional, you've saved the time and cost of one trip.

Combination Screwdriver

A screwdriver that can handle both Phillips and slot screws, plus blades to fit any cowling fastenings. Much of the cost of airplane or avionics repair is in disassembling and diagnosis. Even if you do nothing more than remove access covers, and do rudimentary troubleshooting, you've saved a lot of expense.

Needle Nose Pliers

These can be used for many allowable tasks, such as fishing out something that falls into an inaccessible hole, or making a jury-rig knob if you lose one.

Box/Open End Wrenches

Sometimes a quarter turn on a nut is all that is needed to remake a contact. If you're in a remote area, know what you're doing, and see that a nut is loose, you'll have the right tool to tighten it.

Spray-on Lubricant

Anything from a balky door lock to a squeaky hinge or sticky seat slide can be cured with a spritz of one of these sprays.

Safety Wire and a Sidecutter

Since almost everything on an airplane is safety wired, it's handy to be able to replace or repair a safety that has loosened.

Paper Work

If the flight will take you across national borders, it's a great help if the Right Seat fills out all the customs and immigration forms. In any case, as Bruce Jenner points out, handling and organizing all the paper work around the cockpit is one of the most useful things a Right Seater can do. Many a Right Seat spouse soon takes over the job of filing charts and revisions at home, so the pilot's flight bag is always up to date. That makes her doubly useful in the Right Seat because she knows exactly where everything is filed.

Photography

Whether it's for fun or profit, aerial photography calls for a partnership between Right and Left Seats, with the photographer supplying the artist's eye and technical camera know-how, and the pilot figuring out how to reach and maintain the best vantage point. Aerial shots can be useful to lawyers, realtors, competition bass fishermen, wilderness campers, and many others. And, as artistic expression, photographing crop sculpture is a popular and rewarding pastime.

Survival

Any Right Seater who has special skills in survival, lifesaving, or first aid — and this can include teenagers — can confer with the pilot about what a basic survival or medical kit should contain, depending on the type of airplane, en route terrain, and destination, and on the type of survival or lifesaving training the person has had.

If you know how to set up a solar still, make a tracheostomy, or patch a broken denture, you may as well tuck the basic tools of your trade into a seat pocket.

Small items that would be useless to others could be, in your hands, lifesavers. We've known 12-year-old kids on tropic islands who could make a decent shelter, a sun hat, dinner, and a plate to eat it from, in half a day, using only a pocket knife and whatever they could find in the bush.

Travel Writing

If you're a writer, the airplane becomes your magic carpet. Gordon left a career in corporate aviation because we wanted to become full-time travelers. We lived aboard a boat, explored in our motorhome, and eventually added a Cessna 310 to our stable — supporting ourselves by taking pictures and writing books and articles about our travels.

For any writer in the Right Seat, there is inspiration for fiction or poetry, or for nonfiction having to do with aviation in particular or travel in general. If you're a technical writer, branch out into aviation related topics. Or, take a role as a secretary, reporter, or publicist in aviation organizations and events.

For the writer who writes to sell, the most useful tool is an annual hardcover book, "Writer's Market". It's published by

Writer's Digest Books and is available in any book store. It lists publications by subject matter (e.g. agriculture, travel, women's, bridal, etc.) Also listed are the rules for submission, and the rates they will pay. Book publishers are listed alphabetically, with information about how and what to submit to them. Magazine listings also tell what kind of photos are needed by the publication; if you have *only* photos to sell, get a companion volume titled "Photographer's Market". It too is published annually. So is "Artists' Market", which will help market aviation art.

Volunteerism

Sharp eyes and a willing heart are all you need to volunteer to search for missing aircraft, a lost child, or a boat that did not close its float plan. When a call goes out for help, Right Seat lookouts are needed as much as Left Seat pilots.

If you have specialized skills in the medical or dental fields, missionary work, archaeology, or any of the natural sciences, combine them with your pilot's flying prowess to get involved with any of the many volunteer projects that take place in areas that are accessible only by small plane. You might do a wildlife survey, immunize all the children in a remote Central American village, help in an archaeological dig, or save the rain forests.

How do you connect? Contact wildlife agencies, the appropriate agency at your church, relief agencies of all kinds, the Civil Air Patrol, Explorers, civic and service clubs, and government agencies. Find out whether any of the professional organizations you belong to have volunteer chapters that could utilize the airplane as well as your skills as a doctor, nurse, civil engineer, teacher, or clergyman.

Volunteerism
One Woman's Story

Because Wanda Whitsitt of Champaign, Illinois wasn't satisfied simply to fly without a destination in mind, dozens of downstate Illinois people have been helped.

She was 48 years old and a non-pilot when she joined her husband in taking flying lessons, and she fell head over heels for the flying life.

When she saw a need for means of getting patients from small Midwestern communities to big centers where special, lifegiving treatment was available for their disorders, Lifeline Pilots was born.

Now involving more than 300 pilots in ten states, Lifeline is on call 24 hours a day to fly mercy missions. The organization raises its own funds, schedules its own volunteers and trips, and charges nothing for its services.

Similar organizations operate throughout North America. Known as the Air Care Alliance, the group is developing a toll-free switchboard at a number that was not yet available at press time. Call (800) 555-1212 for toll-free Information. Another nationwide network in which volunteer pilots use their time and talents to provide mercy missions is AirLifeLine, (916) 1716 X St., Sacramento, CA 95818.

Chapter 11

Kids Aboard: Piloting for the Pint Sized

Children and small airplanes go wonderfully well together but, to paraphrase a nursery rhyme, "when it's good it's very, very good and when it's bad, it's horrid." The more the Right Seat can insulate the pilot from any restless, ill, frightened, or just plain bratty children aboard, the better.

Dr. Newell O. Pugh, director of the Cancer Center at Methodist Hospital in Indianapolis, is an accomplished pilot with commercial, multi-engine and instructor ratings, and many hours of flying his Cessna 182 with his wife and children aboard. "Sometimes you just have to close your ears to the screaming baby," he advises other parent-pilots.

As far as the piloting parent is concerned, this isn't *Quality Time*. Nor is it the time to give one of the kids a flying lesson. No cargo is more important than those kids; it's a time for the Left Seat to practice heads-up flying at its most uncompromising, even if it means putting the kids in body casts. With some help from the Right Seat, however, children can — and should — have fun too.

The Right Seat can help by answering questions, refereeing fights, providing toys and entertainment, finding barf bags, passing out food and drink, keeping kids cool enough or warm enough, blowing noses, kissing booboos, and providing frequent updates on Are We There Yet?

Traveling with children in private planes differs from airline flying in several ways, many of them good. The bad news is that you probably don't have a restroom, or room to do anything but sit still. For a child, that's a tall order.

On the plus side, you can plan more frequent stops, seeking out small sod airfields where you can picnic and play tag or touch football to work off excess energy. If the plane is your own, you can evolve the perfect equipment list so it's always aboard — first aid supplies, small blankets and pillows for nap time, a toy bag and that special flying teddy bear, favorite snacks, and juices.

The small plane's atmosphere is different from airliners. At low altitudes, air is moister. Children aren't constantly thirsty, and that helps solve the problem of not having a toilet. Little ears are especially sensitive to altitude changes when any infection is present, but in a small plane flown by a deft and considerate pilot, climbs and let-downs can usually be done so subtly that pressure changes aren't as painful as they are in airliners that can change cabin pressure by thousands of feet in an instant.

Getting the Children Involved

There's good news for flying families. The wonderful world of aviation manufacturers, suppliers, and organizations offers a cornucopia of resources for encouraging, entertaining, and enlightening young people. And most of them are free for the asking. The more you can tap into these resources, the more relevant flight becomes to the teenager in the Right Seat or to the Right Seat parent who is responsible for keeping peace in the Back Seat.

Best of all, the great smorgasbord that is aviation offers something to nourish the talents of all children, not just those who are interested in becoming pilots or astronauts.

Dozens of careers and hobbies have a tie-in with the flying that you do as a family: history, fire fighting, poetry, art, engineering, medicine, social studies, math, geography, meteorology, design, food service, astronomy, writing and communication, and mechanics, to name just a few.

AOPA supports a massive education program, including the APPLE (Pilots Participating in Local Education) project and the Fly a Teacher program in which local pilots volunteer to take a teacher for a short flight to provide a better understanding of general aviation.

NASA provides a wealth of teaching materials. Girl Scouts, Boy Scouts, 4-H, and Explorers have aviation programs, and the Civil Air Patrol invites youth participation.

Through the FAA Co-Op program, students can work for the FAA in a long list of fields including aerospace engineering, air traffic control, maintenance mechanics, electronics, math, and management. When their educational and work requirements have been completed, many are offered full-time career positions with the FAA.

The FAA also offers a Stay-in-School project in which disadvantaged or disabled students can work for the FAA up to 20 hours per week during school sessions, and up to 40 hours per week when schools are closed. Work is usually clerical, with light typing, answering the phone, and filing. For the address of your nearest FAA personnel office, write the FAA at 800 Independence Ave. S.W., Washington DC 20591. Then write the individual office for information about the Co-op or Stay-in-School programs.

Teacher Resources

At this writing, the FAA maintains a dozen Teacher Resource Centers around the United States including one at Embry-Riddle Aeronautical University in Daytona Beach, where the center's coordinator, Patricia J. Ryan set up the program that has been used as a model nationwide.

"Hands on activity is very important in the development of a child," she says. "We want to get children involved. If you're not having fun, find something! If the work is too hard to read, put it on audio tape!"

Bubbling with enthusiasm for both kids and aviation, Ryan

finds that children are often good self-starters. If the teacher motivates the child and gets out of the way, a lot of learning will take place at its own pace and level.

Typical of the centers, ERAU'S offers a library of publications on all levels from kindergarten through 12th grade, plus computers and software and a full range of audio visual materials including videos for use in the home or classroom. "If it's not copyrighted, we'll even dub the video for you," Ryan says. "Just bring in a blank tape."

Although Ryan will work with any interested children and adults who come into the center, much of her material is aimed at teachers in grades K-12. She can provide lesson plans, guest speakers, A-V aids, demonstration materials, study guides, source lists, and much more. By attending her two-week workshop, teachers can gain three credit hours towards their continuing education requirement.

Teacher Resource Centers serving K-12 teachers, parents, and children are found at:

Alabama Technical College, Ozark, AL, (205) 774-5113
University of North Alabama, Florence, AL (205) 760-4623
Embry-Riddle Aeronautical University, Prescott, AZ, (602) 776-3802
National University, San Diego, CA (619) 563-7100
San Jose State University, San Jose, CA (408) 924-3206
U.S. Space Foundation, Colorado Springs, CO (719) 550-1000
Connecticut D.O.T., Wethersfield, CT (203) 566-4417
Embry-Riddle Aeronautical University, Daytona Beach, FL (904) 239-6440
Kansas College of Technology, Salina, KS (913) 825-0275
Parks College of St. Louis University, Cahokia, IL (618) 453-8821
State of Illinois Division of Aeronautics, Springfield, IL (217) 785-8516
Oakland University, Rochester, MI (313) 370-4418
New Hampshire D.O.T., Concord, NH (603) 271-2551
Middle Tennessee State University, Murfreesboro, TN (615) 898-2788
Salem-Telkyo University, Salem, WV (304) 782-5234
EAA Aviation Center, Oshkosh, WI (414) 426-4800

Here are some ways flying parents can involve the children.

At Home

- Join the child in watching aviation-related videos. They're available for all age and educational levels up to and including advanced pilot training courses, for many different areas of aviation interest including documentaries about aviation history, armchair travel, winter flying with bush pilots in Alaska, using airplanes to fight forest fires, and how agriculture benefits from crop dusting.
- Supply reading and activity books having to do with aviation or space. Ask professional librarians for help in finding just the right books for every age and interest level, from "my first airplane ride" to biography, true adventure, and romance novels in which pilots fall in love.
- Let children help in planning a trip. Spread the kitchen table with maps they can understand. Make lists of in-flight snacks, clothes, tapes for the Walkman, or toys to pack, and make a list of the top sightseeing treats in each area. The kid who is a math whiz can work out sample budgets including projected fuel needs; the teenager with the telephone growing out of one ear can phone motels to make reservations or compare rates.
- Fly kites or model airplanes.

At School

- Career day, with visits from professional pilots, mechanics, air traffic controllers, and so on.
- Art or science projects, using aviation materials that are available free or at very low cost from the sources listed in this book.
- Field trips to the airport, nearest NASA center, or science museum. Participate in AOPA's Fly a Teacher program.
- Participation in aviation essay contests, art contents, and science fairs. An annual International Aviation Art Contest is sponsored by the National Association of State Aviation Officials; the Civil Air Patrol sponsors essay contests; the General Aviation Manufacturers Association offers awards to teachers who make an important contribution to aviation education. See addresses in the list below.

Students can compete for scholarships in the FAA's

International Science and Engineering Fair in five areas: behavioral and social sciences, engineering, medicine and health, environmental sciences, and computer science. There is also a grand prize. For information and abstracts, write Science Service Inc., 1719 North St. N.W., Washington, DC 20036.

During the Flight

• Read aviation or travel books related to the flight or destination. The Government Printing Office is a superb source of free and low-cost books including, on the elementary level, "How We Made the First Flight" in Orville Wright's own words, and the "August Martin Activities Book", with games and stories having to do with the nation's first black airline pilot.

• Enjoy a new coloring or activity book.

• Have a "treasure hunt" looking for a list of landmarks you've made ahead of time. One or more can play.

• Put an aviation twist on all the usual games that families resort to when there is nothing to do and no room to do it in — word games using aviation terms, tic tac toe, I Spy, and Hangman. John Ellis, M.D., who is acting medical director in pediatrics at Methodist Hospital in Indianapolis, finds that Game Boy and Nintendo keep their fans occupied for happy hours. "Get an ear phone for the Nintendo if necessary for the user to hear the audio cues," he suggests.

Some kid-care tips for the Right Seater:

• We are all becoming more aware of the importance of wearing ear protection in very noisy environments. Philadelphia attorney-pilot Arthur Hankin, who has flown his family for hundreds of miles, has purchased protective earwear for the whole family. "The kids can listen to their own earphones under the earmuffs," he reports. "They have their own stereos, and we also have an intercom for everyone aboard."

• If one child's fear of flying keeps you from enjoy flying as a family, look into self-hypnosis or relaxation training, suggests Dr. Ellis. Your pediatrician can recommend a qualified therapist. Ellis says that children as young as 2 can be diverted from some

of their fears; children as young as 3 could respond to some forms of assisted relaxation training such as visual imagery.
- The best drinks to carry are sealed paper cartons with a tiny hole for the straw. They are virtually spill-proof.
- Individual boxes of cereal make healthful snacks and minimal mess. Fingers will get less sticky if you favor uncoated types, such as Cheerios.
- Avoid pulltop lids. They're as sharp as razor blades. Even if the child can open and consume the snack or soda without bloodletting, the sharp lids cut holes in your trash bags. And since the airplane's trash bags may sometimes contain pretty disgusting stuff, it's best to protect them from holing.
- Lollipops can provide hours of fun, but give them only to children who are old enough to wield them without getting goo in somebody's hair or the upholstery.
- Salty snacks such as pretzels or saltines help quiet queasy stomachs, but avoid high-fat nibbles such as greasy potato chips or popcorn drenched in coconut oil. Minty hard candies work for some people; candied ginger for others.
- Shop in toy, camping or boating supply stores for motion-proof board games for children. Magnetic types are not a good choice (the compass could be affected), but you'll find many other toys, games and contests that are sealed, contained, or otherwise sticky (e.g. Velcro) for use in solo or team play in a car or airplane. Arthur Hankin warns against buying "talking" toys that can't be heard in a noisy cabin. When the child pulls the string and can't hear the cow say moo, it's no fun at all.
- Always have a change of clothes on board for each child, even if it's only a day trip, and plastic bags in which to seal used diapers, or any clothes soiled by urine or vomit. In a small cabin, bad smells can make everyone sick.
- Although portable potties are available, we just take a couple of coffee cans or other large, disposable containers with snap-on plastic kids. Sometimes it's impossible to get children to use the toilet on demand. Even though a long flight is ahead, nothing happens. Then you take off and..... Sound familiar? If you're trying to get children to use the bathroom before the flight, it sometimes helps to run water noisily in the sink, or flush the toilet.

Dr. Pugh remembers having to make an extra pit stop during a flight with his small children. It turned out to be one of the highlights of their family flying memories. "At the field where we landed, some sort of festival was going on. After we drained the kids, we stayed on and had a wonderful afternoon. It was a good lesson in taking your time. Enjoyment is the name of the game."

- You can't have too many moist towelettes, paper towels, plastic trash bags, and facial tissues.
- The best disposable, waterproof, leakproof bags we've found for urine or vomit are Convenience Bags, sold through aviation catalogs and FBOs. Buy them by the dozen.
- Wring out several clean wash cloths in plain water or in water with a little lemon juice added, and seal them individually in zip-top plastic bags. They're perfect for mopping up little hands and faces.
- If a small child travels in a car seat, look for one labeled "certified for aircraft," a designation that began appearing in the 1980's. Find it on the bottom or back of the car seat, after the phrase "conforms to all applicable federal motor vehicle safety conditions."

Sold through camping suppliers, self-inflating pillows like this one, plus a small blanket, should be kept on board for sleepy children.
(Photo courtesy of the Jerry Martin Company)

- Seat belts are the rule, and it's non-negotiable. We insist that a belt be in place even if the child is lying down, asleep on the back seat. This can be done loosely and in comfort. Don't ever belt a child in with you.
- Inflatable travel "neck" pillows are great for all ages. If you don't have one aboard, seal a bubble of air in a zip-top bag. Slip it into a towel, shirt, sweater or other soft and cuddly garment, and tuck it under a drooping little head.
- Make sure the kids are not too cold or too hot, even if you're perfectly comfortable. Small airplanes always seem to have hotspots and cold drafts hitting individual seats.
- If your children nod off to sleep more easily in the car than in any other spot, it's likely they'll be lulled to sleep by flying too. You may find that the best times to travel are at night and during nap hours.

Aviation Education Resources

When writing to the following groups, specify the child's age and area of aviation interest. Some sources offer materials free, some at moderate cost. Some prefer that requests be made by teachers, on the school's letterhead.

Academy of Model Aeronautics
1810 Samuel Morse Dr.
Reston, VA 22090
(703) 435-0750
Information on building and flying model airplanes

Aerospace Industries Association
1250 Eye St. N.W.
Washington, DC 20005
(202) 371-8400
Information on aerospace manufacturing

Air Line Pilots Association
535 Herndon Parkway
Herndon, VA 22070
(703) 689-2270
Safety, educational, and flying career information

Air Traffic Control Association
220 N. 14th St. #410
Arlington, VA 22201
(202) 522-5717
Information on how ATC works, and on ATC careers

Air Transport Association
1709 New York Ave. N.W.
Washington, DC 20006
(202) 626-4000
Information on the airline industry

Aircraft Electronics Association
Box 1981
Independence, MO 64055
(816) 373-6565
Information on avionics installation

Aircraft Owners and Pilots Association (AOPA)
421 Aviation Way
Frederick, MD 21701
(301) 695-2000
One of the best sources for authoritative information about general aviation. Excellent educational, public affairs, and community involvement programs.

American Institute of Aeronautics and Astronautics
370 L'Enfant Promenade S.W.
Washington, DC 20024
(202) 646-7400
Education materials on aeronautics and astronautics

Astronomical Society of the Pacific
390 Ashton Ave.
San Francisco, CA 94112
Teachers in grades 3-12 who submit a written request on the school's letterhead can get a free subscription to an astronomy education newsletter, "The Universe in the Classroom". Indicate grade level. Teachers need have only an interest in teaching

more about astronomy; a background in astronomy or science is not necessary.

Aviation and Space Education Newsletter
1000 Connecticut Ave. N.W. #9
Washington, DC 20036
(202) 822-4600
This monthly newsletter reports on people and programs in aviation education.

Aviation Distributors and Manufacturers Association.
l900 Arch St.
Philadelphia, PA 19103
(215) 564-3484
Information on aviation products and careers

Aviation Maintenance Foundation Inc.
Box 2826
Redmond, WA 98073
(206) 828-3917
Information, books and technical materials to guide young people who are seeking aviation maintenance careers

Aviation Technical Education Council
2090 Wexford Ft.
Harrisburg, PA 17112
(717) 540-7121
Vocational information on aviation technical careers

Beech Aircraft Corporation
Box 85
Wichita, KS 67201
(316) 676-8839
Aviation education materials

Boy Scouts of America, Aviation Exploring Div.
1325 Walnut Hill Lane
Irving, TX 75038-3096
(214) 580-2427
Write for the address of the Aviation Explorer troop nearest you

Cessna Aircraft Company
Box 7704
Wichita, KS 67277
(316) 946-6192

Aviation education materials including a dandy little plastic model airplane with moveable control surfaces. Teacher's guides and posters, cloud chart, booklet "The Flying Bug" and more, all at very low prices. Request an order blank. Cessna also provides a kit designed to aid Scouts in getting an aviation badge.

Civil Air Patrol
Bldg. 714
Maxwell AFB, AL 36112-5572

Aviation and aerospace education programs for youths aged 13-21

Embry-Riddle Aeronautical University
600 S. Clyde Morris Blvd.
Att: Teacher Resource Center
Daytona Beach, FL 32114
(904) 239-6499

Aviation education materials, programs, teacher support, guest speakers, serving grades K-12.

Experimental Aircraft Association
Wittman Field
Oshkosh, WI 54903-3086
(414) 426-4800

Information on sport aviation, building and restoring airplanes

Federal Aviation Administration
Aviation Education APA-100
800 Independence Ave. S.W.
Washington, DC 20591
(202) 267-3465

Request a list of available materials specific to age group

Contact regional FAA Aviation Education Offices at:
Anchorage, AK (907) 271-5169 Serving Alaska

Atlanta, GA (404) 763-7202 Serving Alabama, Florida, Georgia, Kentucky, Mississippi, North Carolina, Tennessee, Puerto Rico, and the U.S. Virgin Islands

Atlantic City (609) 484-6681 Human resources

Burlington, MA (617) 273-7247 Serving New England

Des Plaines, IL (312) 694-7042 Serving Great Lakes states

Fort Worth, TX (817) 624-5804 Serving Arkansas, Louisiana, New Mexico, Oklahoma, and Texas

Kansas City, MO (816) 426-5449 Serving Iowa, Kansas, Missouri and Nebraska

Los Angeles, CA (213) 297-1431 Serving Arizona, California, Nevada, and Hawaii

Oklahoma City (405) 680-7500 Aeronautical center

Palm Coast, FL (904) 445-6381

Renton, WA (206) 227-2804 Serving Colorado, Idaho, Montana, Oregon, Utah, Washington and Wyoming

FAA education information is also available via computer and modem through FEDIX. Data: (301) 258-0953; Help line (301) 975-0103. Also available from the Washington DC address above is booklet APA-5-149-85, "Teacher's Guide to Aviation Education Resources".

Future Aviation Professionals of America
4959 Massachusetts Blvd.
Atlanta, GA 30337
(800) 538-5627

Career information for advanced high school students who are planning aviation careers. Write for a catalogue listing materials available for purchase.

General Aviation Manufacturers Association. (GAMA)
1400 K. St. N.W. #801
Washington, DC 20005
(202) 393-1500

Information on learning to fly, materials for teachers and schools, general aviation information. GAMA sponsors an annual Learn to Fly Award for Excellence in Aviation Education, for teachers in K-3,4-6,7-9 and 10-12. Prizes include an award certificate, a logbook, and a discovery flight for the teacher or someone chosen by the teacher. Write GAMA for a list of former winners' classroom ideas to put to work in your own family or to share with your child's teacher.

Helicopter Association International
1619 Duke St.
Alexandria, VA 22314-3406
(703) 683-4646
General information on helicopters

Illinois Division of Aeronautics
Bureau of Aviation Education and Safety
Capitol Airport
Springfield, IL 62707
(217) 785-5979

Ask about the Air Bear program, which is aimed at teaching children ages 4-7 about airplanes and calming their fears about flying.

Jeppesen Sanderson
55 Inverness Dr. E.
Englewood, CO 80112
(303) 799-9090
Textbook, videos, classroom aids

Midwest Products Co. Inc., School Division
400 S. Indiana St. or Box 564
Hobart, IN 46342
(800) 348-3497

Request a free catalogue of hands-on aviation activities including balsa model airplane kits and supplies, balsa wood, basswood, tools, and brass tubing or sheet.

Mr. Rogers Neighborhood
Family Communications Inc.
4802 Fifth Ave.
Pittsburgh, PA 15213

Seen on some 300 Public Television stations, Mr. Rogers is a real friend to children in 7 million living rooms. His books deal with everyday situations, including *Going on an Airplane,* and are available in book stores at $7.95. Although the airplane book deals with airline scenes at large airports, it will be reassuring and pleasant reading for all kids who fly. Unlike most children's books, this one is illustrated with photographs and not drawings, so it brims with Mr. Rogers' own, unique credibility.

National Aeronautics and Space Administration
Education Programs Office
400 Maryland Ave. S.W.
Washington, DC 20546
(202) 453-1000

Give your children's ages and ask for a list of available materials for home or school use. Education resource centers are also maintained at NASA centers around the nation. Ask if there is one near you.

NASA's computer at the Marshall Space Flight Center in Huntsville offers its Spacelink, which was developed and donated by Data General Corporation, to educators in any grade level. Any NASA information or educational materials on the system can be accessed, and most can be captured for your own computer or printer.

Dial (205) 895-0028, and set the communications system for 8 data bits, no parity, and 1 stop bit. It handles 300, 1200, or 2400 baud. The call is not toll free. Instructions for logging on will appear on the screen. Press Return, and you'll be asked for a Username and Password. First-timers enter Newuser as both Username and Password, and you're on your way.

The system is packed with information including current NASA news, current and past aeronautic research, the history of the U.S. space program, comprehensive material on the Space Shuttle, an overview of NASA research center, a list of all educational programs including teacher workshops and science fairs, reports on space program spinoffs, and lots of classroom materials. Before signing off, you'll be given the option of leaving of message of up to 15 lines for NASA to read.

National Agricultural Aviation Association.
1005 E St. S.E.
Washington, DC 20003
(202) 546-5722
Ask what materials they have to interest a young person who wants to become an ag pilot

National Air and Space Museum
Office of Education
Washington, DC 20560
(202) 786-2106
Aviation and space information for all age groups

4-H Aerospace Education Program
USDA Extension Service, #3860 South Bldg.
Washington, DC 20250-0900
(202) 447-5516
Ask where to find your nearest 4-H aerospace chapter

National Association of State Aviation Officials
Metro Plaza One
8401 Colesville Rd. #505
Silver Spring, MD 20910
(301) 588-1286
Aviation education materials for all ages and interests. Ask about the annual aviation art contest.

National Intercollegiate Flying Association.
Box 3207 Delta State University
Cleveland, MS 38733
(601) 846-4205
Write for help in forming a chapter at at your local college

The Ninety-Nines Inc.
Box 59965 Will Rogers World Airport
Oklahoma City, OK 73159
(405) 685-7969
Women in aviation contribute to technical and educational efforts

Soaring Society of America Inc.
Box E
Hobbs, NM 88241
(505) 392-1177
Information on gliding and sailplanes

Young Astronaut Council
1211 Connecticut Ave. N.W. #800
Washington, DC 20036
(202) 682-1984
Write and ask how to reach your local chapter.

Audio Tapes Save the Day

Audio tapes, some with accompanying picture books or workbooks, are available for children of all ages. Look in toy stores and book stores. The choice ranges from games and stories to language and education tapes. A large choice of both fiction and nonfiction books is also available for adults.

For children ages 3-8, we recommend *Grandpa Wes' Nature Notebooks*. Read in a soothing, very gentle tone by educator and retired environmental engineer Wes Kunkel, the tapes are gems of entertainment and each carries a science message as well. Little Back Seaters will especially enjoy the story of how the Teeny Tiny Man flies to Mexico in the co-pilot's shirt pocket.

For ordering information about the Grandpa Wes tapes, write Kunkel Artistic Technologies, Box 156, Richland, WA 99352-0156. For a catalog listing language tapes including Spanish and French for children, call Audio-Forum, (800) 243-1234, 9 a.m. to 6 p.m. EST. For other books on tape, including children's classics such as Winnie the Pooh, write Comprehensive Communications, Box 631, Goldens Bridge, NY 10526.

Recommended Reading

Adventuring with Children by Nan and Kevin Jeffrey makes entertaining armchair reading because it is based on a family's firsthand reports on roaming the far corners of the globe with their twin sons, from infancy onward. It is heavy on excellent advice for managing with minimal equipment, and is light enough in tone that it will inspire even the most housebound parents to share the excitement of fly-in adventure travel with their children. The book is published at $14.95 by Foghorn Press/Avalon House, and can be ordered from (800) 842-7477.

Camping with Kids, by Don Wright is a tenting and RV book, but it has plenty of great general camping tips that apply to the fly-in camping family. Wright's advice for keeping children happy in the car applies to the back seat of the airplane too, and his pointers on packing, planning, activities, and campground etiquette are all based on his own camping-with-kids experiences. The book is $9.95 from (800) 955-7373.

Chapter 13

Great Places To Go In Your Airplane

To the flying family, getting there is at least half the fun. With your own private, magic carpet you can fly to mountains and deserts, beaches and dude ranches and campgrounds, cities and the most secret hideaways.

In only an hour or two you can be miles from home, with a whole new horizon and attitude. Work-jangled nerves wind down as you fish, laze by the pool, play golf, set up camp in some wilderness, or check in at a family resort that has activities for everyone from the baby to Gramps.

Here is just a sampling of the many destinations that are ideally suited to the fly-in traveler. Although they have been chosen because of their unique appeal to the general aviation traveler, we have deliberately been skimpy with aviation information about each because it's always best to get the latest information before you go.

Start your own list of places to go for vacations, overnights, and for quick day trips that become special when you go by air. Cincinnati, for example, has a golf course on the airport; a golf course also adjoins Tyler Municipal in Minnesota.

Many airports have destination restaurants, ranging from the "squadron" chain restaurants such as the 94th Aero Squadron at Orlando Executive, to the little lunch counter at the airport in

DeLand, Florida, whose hamburgers were voted among the best in central Florida by readers of the Orlando Sentinel.

Put down at Afton OK, and you're in the Shangri-La resort paradise. Land at Pineville Municipal in Louisiana to watch boat racing championships each fall. There's a boat club right on the field. Try Thacker's airport north of Oil City, LA, where camping facilities are next door to the field. The list goes on and on.

Amelia Island Plantation, Fernandina Beach, FL

For more information: (904) 261-6161 or (800) 874-6878

Price category: Moderate to High

Interest category: Amelia Island Plantation is a total resort for the family vacation, business conference, golf, tennis, sportfishing, and a long Atlantic beach.

Ground transportation: Amelia Island Plantation offers complimentary pickup at Fernandina Beach Municipal. Once you're on the property no car is needed because the the resort-wide

Amelia Island, Florida

The Cockpit Companion

shuttle service is convenient and free. National rental cars are available if you're camping at Fort Clinch State Park or want to stay at one of the historic bed and breakfast inns in Fernandina Beach.

Aviation information: Jacksonville sectional. Paved runways 5000, 5000, and 5350 feet. Full service FBO, fuel. (904) 261-7890.

A century ago one of Florida's major settlements and harbors, Fernandina Beach is today a sleepy backwater town peopled by shrimpers, transient boaters, history buffs, and in-the-know tourists. Summer is high season here. Amelia Island is found in Florida's northeastern notch above Jacksonville, where winter cold fronts can be brutal; so can the bitingly acrid stink of the paper mills when the wind is still or from the mainland.

Nevertheless, 4-star Amelia Island Plantation is one of the state's best tennis and golf centers, often featured in nationally televised, big-name tournaments. The huge resort is a complete, residential community with accommodations of all kinds from entire homes to condos, townhouses, and hotel rooms. Choose yours according to how much space you need, and whether you want to be on the beach, golf course, or tennis courts. Six restaurants, two outdoor bar and grills, 24-hour room service, and two lounges offer plenty of dining options. Since most units (except hotel rooms) have kitchens, guests can also do some or all of their own cooking.

The Bahamas

For more information: After getting aviation information through sources listed below, and deciding on a destination island, get hotel information and reservations through (800) 228-0277.

Price category: Moderate to High

Interest category: Couples, families, singles, casino gambling, deep sea fishing, SCUBA, sailing, tropical romance, nature watching, cultural exchange, history hunting.

Ground transportation: Many resorts are alongside runways, or offer free airport pick-up. Cars and fuel are costly, and driving is on the left, British style, so try to manage without a rental car.

Aviation information: Call the Bahamas Aviation and Sports

Cutlass Bay on Cat Island is one of the many Bahamian out-islands developments that have only a small hotel and an airstrip.

Information Centre, (800) 327-7678 for latest NOTAMS and flight information. An invaluable resource is the Yachtsman's Guide to the Bahamas, published yearly by Tropic Island Publishers Inc., Box 610938, North Miami, FL 33261.

Although we were to spend many years in the Bahamas under sail, our first trip to the islands was in a Beechcraft Bonanza, and we crisscrossed the islands for years by Aero Commander. The very idea of island travel conjures up images of sunshine and tropic nights, wave-washed sands, and an incomparable feeling of hedonistic escape.

What the Bahamas offer that so many other island paradises do not, is proximity to the U.S. mainland. Pop over to Bimini for lunch. It's mere minutes from Miami or Fort Lauderdale, yet it's another world, culture, and nation. Or take off down the necklace of 600 cays, hopping from island to island until you find the Bahama that's right for you.

The list of accommodations ranges from ramshackle hideaways deep in the out islands, to some of the world's most posh resorts.

Planning, confirming, and re-confirming, and a back-up plan for each plan, are crucial to any Bahamas trip, and especially to flights into the outback. Bad weather may have held up the fuel barge supplying the airport; the only resort on the island may have closed; a mechanical problem could mean a long wait until a needed part can be flown in. Bahamas flying is adventuring at its best for those who prepare for it and welcome its challenges.

Chalet Suzanne, Lake Wales, FL

For more information: (813) 676-6011
Price category: High
Interest category: Couples, romantic getaway, gourmet dining.

Ground transportation: Although a courtesy car is available, the runway is only footsteps from the door of your suite. No car is necessary during your stay.

Aviation information: Find it on the Miami sectional. The resort has its own 2,450 X 50 turf runway.

The Chalet has received more press and plaudits for its dining than almost any other restaurant in Florida. Soon after its

Long a mecca for fly-in visitors, Chalet Suzanne near Orlando, FL is operated by aviation pioneers Carl and Vita Hinshaw. The inn, famous for its cuisine, offers lodgings that are only a few footsteps from the runway. (Photo courtesy of Chalet Suzanne)

founding by Bertha Hinshaw during the darkest years of the Depression, Chalet Suzanne was discovered by food writer Duncan Hines, who sang its praises in print.

Now operated by aviation pioneers Carl and Vita Hinshaw, the secluded complex has a pool, lake and lawns, its own pottery and a small antique shop, and a soup cannery where the Chalet's own soups are canned. (Some of them went to the moon with the Apollo astronauts).

The original inn burned during the war when no building materials were available, so the widowed Bertha cobbled together what she could out of old chicken coops and stables, and created a delightfully eclectic hodgepodge as comfortable as an old house slipper. All rooms are different; all have a sort of faded, fairytale gentility.

Dinner is an event. It begins with the Chalet's famous broiled grapefruit, then its unique romaine soup, an equally unique salad, a sorbet intermezzo, a choice of entrees, then crepes followed by dessert. Dinners can cost $60 per person or more depending on the wine; rooms are in the $140 range. The best value, available in off season only, is a package that includes room, dinner, and breakfast. Or, land long enough for lunch, which will cost about $35 per person.

Reservations are always a good idea.

Crested Butte, CO

For more information: (800) 525-4220 and (800) 545-4505.

Price category: Moderate

Interest category: Rugged, fresh air vacation featuring mountain biking and hiking, wildflowers, old mining towns, camping, hang gliding, trout fishing. Couples, families, singles.

Ground transportation: Check with your host to see if pickup is available at the airport. If not, take a cab. Distances are short, and once settled in at most lodgings here, you can manage without a car.

Aviation information: Crested Butte (3V6), elevation 8,908, is found on the Denver sectional. Open only in summer, it has a 4,500-foot asphalt runway. Call (303) 349-7334 for information. At this writing, a $50 landing fee is charged.

Because runways and mountain aeries are often mutually

The Cockpit Companion 147

Crested Butte, CO is a complete mountain playground with an airport that is open in summer.
(Photo courtesy of Crested Butte Resort Area)

exclusive, the vacation smorgasbord offered by the Crested Butte Resort Area, which has its own airport, is indeed a find. Nestled in the valley, and creeping up the mountain, the settlements are connected by a free shuttle bus.

We stayed at the Grande Butte Hotel, which has its own restaurant and is in a little cluster of shops and additional restaurants. If you'd rather do your own cooking, many condos and villas are available in the resort area, most of them bargain priced in summer. (Ski season is high season here.)

The Silver Queen ski lift operates through the summer, providing access to picnicking, hiking, and hang gliding high on the mountain. Your hotel can arrange activities such as fishing, boating and horseback riding. Spend at least a day walking around the village, with its old mining town flavor and its many excellent restaurants. Or just rent a bicycle and take off on mountain roads to smell the flowers.

Downtown Chicago

For more information: The Drake, (312) 787-2200; The Stouffer Riviere, (312) 372-7200; The Omni Ambassador East, (312) 787-7200.

Price category: Moderate to High

Interest category: Business, power shopping, luxurious weekend getaway, theater, big-city sights and lights.

Chicago's Ambassador East Hotel, home of the famous Pump Room and a short cab ride from Chicago's lakefront airport, offers romantic weekend packages. In its Author's Suite are hundreds of books, all of them autographed by famous authors who have stayed here. (Photo courtesy of Omni Hotels)

Ground transportation: Best bet is to take a cab to the hotel, then get around on foot, by city bus, or by cab.

Aviation information: Meigs (CGX), with its 3,948-foot runway, is in the heart of downtown Chicago and in the shadow of McCormick Place. Find it on the Chicago sectional. For information, call (312) 744-4787.

One of the best uses for private planes owned by rural and suburban families who live in Illinois, Indiana, lower Wisconsin and Michigan, and Iowa is to descend on downtown Chicago for a taste of all that a big city has to offer, with none of the traffic hassles.

Shop Marshall Field's, which covers an entire square block, and the legendary stores along the Miracle Mile. Haunt the city's comedy and jazz clubs, or see a big-league game. See some of North America's best theater, opera, dance, museums, zoos, and gardens. From the Drake, you can even walk to a sandy Lake Michigan beach. Sample world class dining too, starting with the celebrated Pump Room at the Ambassador East and ending with the famous Cape Cod Room at The Drake, with a stop at the revered Pizzeria Uno in between.

The Drake is a regal, historic hotel often chosen by visiting royalty. The Ambassador East burst on the scene during the Hollywood glamour years, when Bogie and Bacall honeymooned here and the Pump Room became The place to be seen. It's still a favorite with the likes of Sinatra, Ann Jillian, Oprah Winfrey, and countless others. In the hotel's Authors' Suite is an entire library of books autographed by famous writers who have stayed here. As a new hotel, the Stouffer Riviere offers 27 floors of luxury, located between the Loop and the Miracle Mile overlooking Lake Michigan and the Chicago River.

We've listed the hotels above for two reasons. First because they are among our personal favorites for first class surroundings and service in locations that are handiest to the airport and the best, downtown sites. Second, because they offer discounted weekend packages that allow ordinary folks to live like rajahs for an unforgettable few days before returning to the everyday grind.

Downtown Cleveland

For more information: Convention and Visitors Bureau, 3100 Terminal Tower, Cleveland, OH 44113, (216) 621-4110 or (800) 321-1001.

Price category: Moderate to High

Interest category: Big-league sports, city shopping, dining, concerts, theater, nightlife, attractions. Couples, business, singles, families.

Ground transportation: Free pickup at Lakefront is offered by the lavishly restored Stouffer Tower City Plaza Hotel. Most points of interest including the playhouse district are within walking distance, or a short cab ride, from the hotel. Call (800) HOTELS 1 and ask about weekend Breakation packages.

Aviation information: Burke Lakefront Airport is found on the Detroit sectional. A 24-hour, full-service field, it has paved runways 6,200 and 5,200 feet long. Landing and parking fees are charged. For further information, call the airport at (216) 781-6411, or FBOs at 781-1200 and 861-2030.

Burke Lakefront Airport, like Chicago's Meigs, allows fly-in travelers to plunk down in the inner city without battling hours

Located on the site that has been an inn since 1815, Cleveland's historic Stouffer Tower City Plaza Hotel will pick up fly0in visitors at the airport, using stretch limousines. The service is free.

of suburban traffic. Once here, all the glamour of a major city is at your doorstep. Cleveland, one of those comeback miracles of urban restoration, is a vibrant city that loves the arts, elegant shops, fine dining, and history in addition to its sparkling lakefront with all its watersports. Start with a tour on Lolly the Trolley to get an overview of downtown, the coast, the historic riverfront Flats, the artist colonies, universities, and more.

Then strike out on your own, choosing from a long menu of things to do and see, places to eat, and hot-ticket games or theater.

Fort Scott, KS

For more information: Chamber of Commerce, 231 E. Wall, Fort Scott, KS 66701, (316) 223-3566 or (800) 245-FORT.

Price category: Low to Moderate

Interest category: Historic, military, heartland America, families, couples.

Ground transportation: Call ahead to reserve the FBO's free courtesy car. Once settled in, you can get around town on foot or on the free trolley.

Aviation information: (316) 223-5490. Found 90 miles south

Fort Scott was built in 1842 to patrol the permanent Indian frontier. It is now a fully preserved National Historic Site.
(Photo courtesy of the Fort Scott Area Chamber of Commerce)

of Kansas City, the airport has a 4,500-foot hard surface runway, and is lighted for 24-hour use. The FBO is open 7 a.m. to 7 p.m. daily.

Travel writers overwork words like "charming" and "quaint" but Fort Scott epitomizes them. Although the town dates back to 1842 and the Mexican war, most of the buildings here are vintage Victoriana, now restored into shops, offices, and bed and breakfast inns. Even the "meller-dramas" presented by the local theater are firmly stuck in the good old days.

The fort, which has been restored to its appearance in 1842-1853, has living history demonstrations on summer weekends. In mid-September, fracases from the Mexican war are reenacted. The National Cemetery near here, created by President Lincoln in 1862, is one of the nation's largest.

Come to see the fort, and make a weekend of it by booking a room at the Chenault Mansion (316) 223-6800 or the Huntington House, 223-3644. Both are bed and breakfast inns.

Jekyll Island, GA

For more information: Jekyll Island Convention & Visitors Bureau, Box 3186, Jekyll Island, GA 31520, (800) 841-6586 or (912) 635-3636.

Price category: Low to Moderate

Interest category: Families, business, romantic getaway, golf, tennis, historical, Atlantic beaches, deep sea fishing.

Ground transportation: National Rental Car, or arrange with your hotel for airport pick-up. Once settled in, the best way to get around the island is by rental bike.

Aviation information: (912) 635-2500. Found on the Jacksonville Sectional, 09J has a 3718-foot asphalt runway.

A century ago, America's wealthiest and most powerful moguls turned this island into a private hunting preserve with its own palatial Clubhouse. In time, many of them built Newport-like "cottages" around the central club, eventually adding indoor tennis courts, a clinic, and their own church with Tiffany windows. Now restored, the complex is owned by the state.

Edged to the east by wave-washed beaches and to the west by Georgia's famous "golden" marshes and the Intracoastal

Waterway, the island has a campground, a small cluster of stores where essential supplies can be purchased, ruins of l8th century plantations, and acres of things to do and see. Take the children to the water park. Play golf or tennis. Book a deep sea fishing trip, or drop a line from the free fishing pier. Swim in pools or the ocean. Picnic in the dunes. Explore the millionaire mansions, which have been restored and opened to the public.

A list of accommodations is available from the address above. Both hotel and housekeeping units are available on the beach. The campground is inland, in groves of towering, twisted live oaks. Crown jewel of the hotels is the original, l9th century Clubhouse, now caringly restored and a Radisson resort called the Jekyll Island Club Hotel. It too is inland, overlooking the marshes and the Intracoastal Waterway.

Typical of tropical Victoriana, the hotel has cupolas and verandas, plenty of gingerbread, and lushly groomed grounds including the traditional croquet lawn. The Club has a spacious pool, and free shuttles run regularly to the beach. It's within walking distance to the Millionaire's Village and to the marina, where fishing charters are available.

Monterey Peninsula, CA

For more information: Carmel Innkeepers Association, Box Y, Carmel, CA 93921 or the Pebble Beach Company, Box 567, Pebble Beach, CA 93953.

Price category: Moderate to High

Interest category: Romantic getaway, scenic drives, nature watching including the famous Monarch butterfly return in October and the seal pups along the coast.

Ground transportation: Rental cars are available at the airport. If you don't want a car, book at a resort that offers airport pickup.

Aviation information: Monterey Peninsula (MRY) is found on the San Francisco sectional. Its runways are 6,597 and 4,001 feet, with transmitter activated lights and an ILS. (408) 373-3731.

The breathtaking Big Sur country suggests long drives along the coast, sightseeing at the Carmel Mission (1770) or the aquarium, visits to the Steinbeck House in Salinas and to the Hearst Castle, long walks, shopping in trendy boutiques and Carmel's

The Lone Cypress Tree, growing on the rocky coast at Pebble Beach, is one of the points of interest along picturesque 17-Mile Drive, and is the inspiration of Pebble Beach Company's logo/trademark.

dozens of galleries, fishing, kayaking, golf and tennis, fine dining and great wines. Although we usually choose destinations where a car won't be needed, the famed 17-Mile Drive, from the coastline through the forest, makes driving part of the joy of this place.

Palm Springs, CA

For more information: Palm Springs Tourism, 401 S. Pavilion Way, Palm Springs, CA 92262, (619) 778-8415 or (800) 34-SPRINGS.

Price category: Moderate to High

Interest category: Golf, biking paths, nightlife, jogging trails, theater and entertainment, tennis, swimming, horseback riding in Indian canyons, aerial tramway to ski areas and horse trails, hot springs spa. Families, couples, singles.

Ground transportation: Ask your hotel host. You can probably manage without a rental car. Taxis and rental cars are available, and some resorts offer courtesy airport pickup.

A short flight from the Los Angeles area, the Palm Springs resort area is served by an airport that is only a mile from the heart of downtown. Once here, you can enjoy a long list of activities.
A rental car is not needed.
(Photo courtesy of Palm Springs Tourism)

Aviation information: Palm Springs Regional (PSP) has an asphalt runway 8,500 feet and a VOR approach. It's found on the Los Angeles sectional, and offers full FBO services. For more airport information call (619) 323-8161.

Because the airport is only a mile from the heart of downtown, this desert playground is the perfect destination for the fly-in traveler. Choose from 150 lodgings ranging from simple bungalows and bed and breakfast inns to some of the nation's most swank resorts.

Get around by bicycle, horsedrawn carriage or on foot. Every

Thursday night, there's an old-fashioned street fair downtown. Streets are closed to traffic, and are turned into an international bazaar with music, food, and shopping. Food is a fetish here, especially food for the thin-is-in crowd. Chef Robert Pavese at the Wyndham Palm Springs is a guiding force behind the hotel chain's exciting new nutritional cuisine, featuring low-fat, high fiber, low-salt menus that are high on class, appearance, and taste. Try the Cocoa Chiffon Cake or the Pork Chops with Pear and Ginger.

Sugar Loaf Resort, Cedar, MI

For more information: (800) 748-0117 or (616) 228-5461

Price category: Moderate

Interest category: Families, couples, business conference. Golf, tennis, croquet, badminton, fitness club and classes, walking and biking trails in summer, skiing in winter.

Ground transportation: Rental cars available, but the runway is right on the resort, so a car is not needed unless you want to tour local area wineries or the Sleeping Bear National Lakeshore.

Aviation information: Y04 is found on the Green Bay sectional, 14 miles NW of Traverse City and has a paved, 3700-foot airstrip that is open March through October. (616) 228-5461.

Located on the beautiful Leelanau Peninsula, this popular resort is only 148 miles from Grand Rapids, 244 from Detroit, and 320 from Chicago. Stay in a hotel room, a condo with full kitchen and one or two bedrooms, or a townhouse with up to four bedrooms.

American cuisine is celebrated in the resort's fine dining room, Four Seasons, where Sunday brunch is an epic event. Choose from a variety of eateries located on the resort, do some of your own cooking if you're in a townhouse or condo, or rent a car and try area restaurants. The resort also has an old-fashioned saloon with Mexican food and a game room, and the dressy Top of the Loaf Lounge with dancing, entertainment, cocktails, and a see-forever view of Lake Michigan sunsets.

The Tides Inn, Irvington, VA

For more information: (800) 843-3746 or (804) 438-5000

The Cockpit Companion

Sugar Loaf Resort in Cedar, MI spreads a warm welcome, and a landing field, for fly-in visitors. (Photo courtesy of Sugar Loaf Resort)

Price category: Moderate

Interest category: Families, business conference, honeymoon, golf, boating, tennis, sightseeing nearby historic sites.

Ground transportation: Free pickup at the airport; once at the Inn, no car is needed unless you want to take a sightseeing tour.

Aviation information: Hummel Field (W75) is found 6 miles east of Saluda on the Washington sectional. The VFR, 2500-runway is paved and lighted. For information, call (804) 758-5334. If you need a 5000-foot runway, use West Point, 45 minutes to the west. The Inn will pick you up there for $15.

The serene, waterside setting of tidewater Virginia welcomes fly-in visitors to a gracious resort that offers a complete family vacation or a romantic getaway. Play 45 holes of golf including the famous Golden Eagle course, take a yacht cruise on the Rappahannock River, enjoy beaches and bikes, and dine on the American Plan.

The atmosphere is pleasant and just prim enough (gentlemen wear jackets at dinner), to give an air of elegance that lures insiders back year after year.

The Cockpit Companion 159

INDEX

A&P mechanic 25, 27, 117-118
Abeam 8
Abort 8
Academy of Model Aeronautics 132
Aerospace Industries Association 132
Air Line Pilots Association 132
Air Traffic Control Association 133
Air Transport Association 133
Aircraft Electronics Association 133
Aircraft Owners and Pilots Association 133
Airworthiness Directive 27
ADF 6, 76-77
ADIZ 8
Addresses, aviation education sources 132-141
Aerial photography
AGL 9
AI, authorized inspector 25, 27
Ailerons 35, 38
Aircraft, names of parts 35-41
Airspeed indicator 41, 44-46
Airsickness 93
Air pocket 8
Airway 8, 15, 73
Alphabet, phonetic 16
Alternate air 52, 72
Alternate airport 9, 32
Altimeter 33, 44, 84
Altimeter setting 30, 44
Amelia Island Plantation 142-143
American Institute of Aeronautics and Astronautics 133
Astronomical Society of the Pacific 133
Ammeter 50
Antennas 40, 44, 92
    ADF 796
    Emergency Locator Beacon 41
    glide path 41
    Loran 41
    Omni 40
    sense 41
    transponder 41
    VOR 40
AOPA 9, 25, 125, 127, 133
APU 9, 54
ARTCC 9
Artificial horizon 47
ATC 7, 9, 13, 14, 15, 133
ATIS 9
Avgas 61
Aviation art contest 127
Aviation and Space Education Newsletter 134
Aviation Distributors and Manufacturers Association 134
Aviation Maintenance Foundation Inc. 134

Aviation Technical Education Council 134

Bahamas 143-145
Beech Aircraft Corporation 134
Boy Scouts of America 134
Bulletin, see AD 8

Camping, fly-in 113-114
Carburetor heat 52, 53-54, 72
Cessna Aircraft Company 35
Chalet Suzanne 145-146
Charts 8, 9, 10, 15, 74, 75, 76, 79, 85, 119
    En route 59, 60
    sectional 14, 74
    WAC 59, 60, 74
Checklists 66, 67-70, 84
Chocks 39, 64
CG (Center of Gravity) 10, 38
Charter flying, about 26
Chicago 148-149
Circuit breakers 54
Civil Air Patrol 9, 121, 125, 127, 135
Clear the propeller 39, 68, 94
Clear, clearance 10, 12, 68
Cleveland 150-151
Clouds 16-17, 32, 33, 135
    Alto 17
    Cirrus 17
    Cumulus 17
    nimbus 17
    stratus 17
Collision course, how to recognize 86-87
Compass 10, 46-47, 60, 76, 77, 118, 130
    Gyro 47-49
    Magnetic 46, 47, 49
    Slaved 49
Computers 114-117
Control locks 41, 64, 65, 85
Controls, engine 52-54
Convenience Bags 131
Cowling 39, 64, 118
CPR 8, 102
Crested Butte, CO 146-148
Crosswind 10, 74
Cylinder head temperature gauge 50

dB 10
Directional Gyro, DG 10, 49
DME 10, 15, 41, 78-79
Door, cabin 66, 72
DUATS 114, 115, 116

Elevator 37, 38
Ellis, Dr. John 129

Index

ELT 10, 92, 93
Embry-Riddle Aeronautical University 125, 126, 135
Empennage 35
EPIRB 92, 93
ETA 10
Exhaust gas temperature gauge 50
Experimental Aircraft Association 10, 135

FAA 8, 10, 11, 21, 22, 23, 24, 27, 114, 115, 116, 117
   Co-op program 125
   FAA AME (Medical Examiner) 123
   Education offices 136
   stay-in-school program 125
   FEDIX database 117
FAR 10, 23
FBO 10, 12, 114
FCC 11, 117
Fin 37
Final approach 11, 36
Flaps 35, 36, 37, 46, 72
Flight plan 58, 59, 62, 85, 115
Fog 33, 40
Fort Scott, KS 153
4-H Aerospace Education Program 139
Feet per minute 11, 46, 84
Frequency 9, 11, 12, 40, 76, 77, 78
FSS 11, 114, 115
Fuel 4, 9, 14, 30, 33, 38, 50, 53, 54, 58, 60, 62, 66, 67, 68, 74, 84, 85, 127
   caps 37
   drains 37
   gauge 51
   sump, check 60-61
   vents 37
Fuses 55
Future Aviation Professionals of America 136

General Aviation Manufacturers Association 136
Ground Control, GCA 11, 68
Gear Warning horn 84
General aviation 5, 9, 11, 79, 125, 127, 133, 136, 137, 143
Generator 55, 45, 66, 72
Glare 46
GMT (also see Zulu) 11
Go-around 11, 13, 36
GPS 6, 10, 79, 79-82
Ground speed 12, 46, 79, 84
Gyro instruments 47-50, 51, 69, 77
Gyro, sandwich 47
Gear 39, 70, 73
   fixed 40
   main landing 40
   retractable 40

Hankin, Arthur 94
Heat, cabin 51-52

Helicopter 26
Helicopter Association International 137
Holding pattern 12, 13
Homing beacon 12, 13
Hour meter 51
Hub (propeller) 39, 69

IFR 12, 32, 33-34, 48, 58, 74, 76
Idle cutoff 53
Illinois Division of Aeronatucs 137
ILS 12, 79
Inspections, aircraft 8, 27, 60, 64
International Airports 12
Intoxication, effects of 89
Instrument panel 6, 43-55, 77, 101

Jekyll Island, GA 152-153
Jenner, Bruce 20, 64, 66, 72, 103, 119
Jeppesen Sanderson 137

Knots 12, 41

Life vest 91-93
Life raft 91-93
Lights 73
   cabin 85
   rotating beacon 40
   navigation 40, 86
   strobe 40
Loran 41, 79, 81, 82

Magneto 55, 66, 99
Master switch 55, 118
Main (landing) gear 40
Mayday 12, 44
Mechanic, A&P 19, 25, 27, 61, 117
Mechanics, amateur 117-119
Menu suggestions 106-110
Midwest Products Co., School Division 137
Minimums 11, 32, 34
Missed approach 11, 13
Mr. Rogers Neighborhood 138
Mixture control 45, 50, 52, 54, 68, 70
Monterey Peninsula 153-154
Multi-meter 118

NASA 138
National Agricultural Aviation Association 139
National Air and Space Museum 139
National Association of State Aviation Officials 139
National Intercollegiate Flying Association 139
Navaids 75
Nautical mile 12, 78
Navigation lights 40, 86
Night flying 85
Ninety-Nines 140
Nose gear 39, 40, 103

The Cockpit Companion 161

NOTAM 11, 13

Oil pressure gauge 45, 50
Oil temperature gauge 45, 50

PCA 13
Packing 110-111
Palm Springs, CA 154-156
Pets on board 90
Pilot, Private 22, 23
Pilot, Student 21
Pilot qualifications 20-25
Pinch Hitter (course) 28
PIREPS 7, 13
Pitot tube 13, 41, 45, 64
Port 8, 14
Precipitation, precip 14, 40, 74, 83
Propeller 5, 37, 52, 53, 45, 68, 73, 85, 69, 70, 94-95
    Constant speed 5, 53, 68
    Controllable 6, 53, 68, 73
    Propping, how to 99-100
PTT switch 14, 71
PTR switch 71
Pugh, Dr. Newell 123, 131

Ryan, Patricia J. 125, 126
Radio
    communications 16, 26, 30, 70, 71, 72
    ham 117
Ramp 14, 60
Rate of Climb (ROC) 11, 46
Ratings 20, 21
    Airline Transport Pilot 22
    CFI (Instructor) 22
    Commercial 22
    Instructor 22
    Instrument 22
    Multi-engine 23
    Other 23
    Private 22
Red line 46, 50
Retractable gear 40
Rotating Beacon 40
Rudder 37, 38, 41, 69, 75

Science fairs 127, 138
Scouts, boy, girl, explorer 125, 134, 135
Seat belts 70, 72, 131
SIGMET 14
Slats 35
Soaring Society of America Inc. 140
Spars 37
Speed brakes 37
Spinner 35, 39
Spoilers 37
Stabilizers 37

Stall 14
    stall warning horn 84
Starboard 8, 14
Statute mile 14
Strobe 40, 55, 86
Struts 37
Suction indicator 45, 50
Sugar Loaf Resort 156
Survival courses 91-93

TAC 13
Tachometer 45, 50, 53, 73
Tail skid 40
Takeoff 36, 66, 69, 70-73, 84, 90, 92
TCA 15
Teacher Resource Centers 125-126
Telltales 40
Throttle 39, 45, 52-53, 72, 73, 99
Tides Inn 156-158
Tiedown ropes 64
Tires 64
Tools, suggested list 117-119
Touch and go 15
Trailing edge 35, 37, 38
Travel writing 120-121
Tricycle gear 39
Trim tabs 38, 45, 52, 69, 72
Turn indicator, turn and bank 45, 49
Tyler Municipal 143

Unicom 71, 84

Vacuum gauge 50-51, 69
Vector 15
Vents, air 52
Vents, fuel 36
Vertical Velocity Indicator 46
Vertigo 47
Victor15, 16, 59, 115
VFR 4, 15, 32, 45, 58, 59, 74, 75, 76, 84, 115
Voltage meter 50
Volunteerism 121-122
VOR, VORTAC 13, 15, 40, 41, 77-78

Weather 11, 117
    briefing 30, 32, 33, 34, 57, 114, 115
    clouds, definitions 16-17

Yoke 15, 52, 71, 69
Young Astronaut Council 140